SKIN WISE

A No-Nonsense Guide to Skincare Ingredients

BARBARA ATKINSON

Disclaimer

This book is for educational and informational purposes only. It is not intended as medical advice and does not substitute for individualized diagnosis, treatment, or professional guidance. Always consult a qualified skincare or healthcare professional for questions about your skin, health conditions, or product use. The author and publisher disclaim any liability arising directly or indirectly from the use of the information contained herein.

Trademarks

All trademarks, service marks, product names, and company names are the property of their respective owners and are used for identification purposes only. Use of these names does not imply endorsement.

ISBN: 979-8-9930680-0-8 (eBook)
ISBN: 979-8-9930680-1-5 (Paperback)

First Edition

www.TheSkinWiseBook.com

Table of Contents

INTRODUCTION

Great looking skin is healthy skin, and what you put on it matters. That's not marketing hype; it is fact. The ability to read and understand skincare ingredient labels is just as important as knowing what's in your salad dressing. The big difference? Your skincare choices are far more likely to show on your face.

In an industry overflowing with buzzwords and exaggerated claims, the ingredient label tells the real story. It reveals active ingredients, their relative concentrations, which ingredients serve as fillers, and which may be controversial.

Becoming fluent in skincare ingredients means understanding the label language, how ingredients function, how they interact with each other, and most importantly, how they interact with your skin. Learning how to do this with confidence is easier than you may think.

SKIN WISE has been designed to be your guide to smarter, more effective skincare. It's organized into two parts:

PART I: HOW YOUR SKIN WORKS & WHAT IT NEEDS is designed to equip you to navigate the skincare aisle with confidence.

- **Know** the different layers of your skin and how they work together
- **Understand** how ingredients interact with your skin and each other
- **Identify** the types of ingredients and regimen your skin currently needs
- **Recognize** skincare marketing tactics and gimmicks
- **Understand** what skincare ingredient labels are telling you

PART II: THE SKINCARE INGREDIENT REFERENCE is a practical, science-based mini-encyclopedia of skincare ingredients. Becoming fluent in skincare ingredient labels doesn't require memorizing long and difficult to pronounce ingredient names.

The ingredient reference is organized by functional groups - cleansing surfactants, toner, hydrators, targeted actives, and so on - to provide you with an understanding of how the different ingredient categories function. The reference tells you how they're commonly paired, how their roles shift across product types, their skin compatibility, and what they mean for your skincare strategy. The ingredients detailed for each functional group are those you are most likely to find in skincare products on the shelf.

This contextual layer helps you understand not just what an ingredient is, but what it means in the formulation. Instead of just decoding ingredients one-by-one, you'll understand why they're arranged that way, what the formula is trying to achieve, whether it's well-designed, and how it compares to similar products.

This foundational knowledge empowers you to easily recognize which ingredients are active, which are functional, which are simply there for texture, scent, or shelf stability, and most importantly, whether it's right for your skin.

When you use skincare ingredient pattern recognition in **SKIN WISE**, you'll know that niacinamide is effective for strengthening skin barrier, controlling oil, brightening, and reducing redness. You'll also know that an effective concentration is 2% to 10%, so it should appear in the 5th to 8th position in a serum ingredient list.

SKIN WISE is not a replacement for new apps like the INCI Decoder, which can tell you that niacinamide is a skin-restoring ingredient. It is a companion. designed to help you quickly and effectively identify what your skin needs and how to determine whether a particular product is right for you.

Welcome to your resource for more informed, more effective skincare choices!

PART I: HOW YOUR SKIN WORKS & WHAT IT NEEDS

Your skin is not simply a passive surface. It's a dynamic, multi-layered organ designed to protect against environmental threats, regulate hydration, and continuously repair itself. Skincare ingredients interact with these layers in distinct ways, sometimes not doing much, sometimes working only on the surface, and other times penetrating deeper to create long-term change.

The effectiveness of a skincare formula depends on several factors: molecular size, solubility, concentration, the condition of the skin barrier, and delivery method. This opening section of **SKIN WISE** provides:

- An understanding of your skin's functional foundation, including its barrier and regenerative systems

- A breakdown of how different ingredients interact with the various skin layers of your skin, from surface hydration to deeper repair

- A framework for identifying which ingredients your unique skin needs to address specific concerns and optimize skin health

- Insight into the marketing tactics and claims that misrepresent product effectiveness or appropriateness

- The skills to interpret product label language, empowering confident and informed product choices

By the end of **PART I**, you will be fluent in the fundamentals of skin science, skincare ingredients, and label literacy. You will be equipped to approach skincare with the same confidence you bring to your nutrition or fitness choices. This foundational knowledge makes the ingredient entries in **PART II - THE SKINCARE INGREDIENT REFERENCE** not only informative, but actionable.

HOW YOUR SKIN WORKS

Before you apply a single serum or invest in another skincare product, you'll make better decisions when you have a better understanding of how your skin functions. Your skin isn't just a surface; it's a living, layered system performing complex tasks 24/7. Every glow, breakout, wrinkle, or irritation reflects what's happening beneath the surface.

In this chapter, you'll get a science-backed breakdown of how your skin is built, what each layer is responsible for, how those layers work together to protect and renew you, and how skincare ingredients can affect each layer's function.

With this foundation, you'll understand how the skin barrier functions, how regeneration happens, and how damage takes hold. This is the knowledge that will help empower you to choose ingredients and treatments that strengthen your skin, rather than just reacting to symptoms.

Let's start from the top, literally, and move step by step through each layer, uncovering how structure, function, and targeted care all come together to create the skin you see in the mirror.

Skin Basics: How Your Skin Is Built

Your skin operates through two fundamental mechanisms that determine most everything from how young you look to how well your products work: **cell turnover** that constantly renews your skin's surface, and **skin barrier function** that protects you from the outside world while keeping moisture locked in.

When these fundamental processes work optimally, your skin looks smooth, feels comfortable, and responds well to the products you use. But when turnover slows or your barrier becomes compromised, which happens more easily than you might think, skin dullness, irritation, dehydration, and accelerated skin aging follows.

Your skin barrier is more than just a simple shield; it's a layered ecosystem. Each layer contributes to barrier integrity, hydration, immunity, and communication. Here's what you need to know.

Stratum Corneum: Your Shield Against the World

The stratum corneum is the outermost layer of your skin. You can think of this layer as a brick wall. The "bricks" are flattened dead skin cell flakes (corneocytes) and the "mortar" is made of ceramides (a type of fat), cholesterol, and fatty acids.

This outer skin barrier serves two critical functions. First, this semi-permeable layer restricts entry of harmful substances and blocks pathogens. Equally important, it minimizes Transepidermal Water Loss (TEWL), meaning the evaporation of water from deeper skin layers.

When your stratum corneum is compromised by harsh cleansers, over-exfoliation, or stress, your skin feels tight, looks dull, and becomes sensitive and reactive. Repairing a compromised stratum corneum layer, covered in detail in later chapters, requires gentle cleansing, moisturizers rich in ceramides and natural moisturizing factors (NMFs), and barrier-focused care.

The Living Epidermis: Where Fresh Skin is Made

The layers beneath the stratum corneum are alive, constantly regenerating to maintain your skin barrier:

- **Stratum Lucidum**: Found only on palms and soles, adds protection in high-friction areas.
- **Stratum Granulosum**: Produces lipid-rich lamellar bodies that waterproof your skin.
- **Stratum Spinosum**: Strengthens skin with keratin and desmosomes; houses immune Langerhans cells.
- **Stratum Basale**: The root of renewal. Basal cells divide and migrate upward while melanocytes provide pigment and UV protection.

While this overview gives you the basic framework of your epidermis, understanding how each layer functions, and what happens when these functions are disrupted is important for choosing the skincare ingredients your skin needs to be healthy and glow.

Each layer has specialized roles that determine not only how your skin looks and feels today, but also how it will age and respond to the products you use. Let's examine each layer in detail to understand exactly what's happening in your skin and how to support optimal function at every level.

Stratum Lucidum: Extra Protection for Tough Spots

The stratum lucidum is a thin, transparent layer of skin unique to the thick skin found on the palms of the hands and soles of the feet. This layer consists of flattened, dead keratinocytes containing eleidin, a protein that contributes to its clear appearance. The stratum lucidum's primary function is to provide an additional barrier to water loss and protect the underlying skin layers on your palms and bottom of your feet.

Stratum Granulosum: Building a Strong Barrier

The stratum granulosum, also known as the granular layer, is essential for forming the skin's barrier's waterproofing ability. This layer is relatively thin compared to others, typically just 1 to 2 cells thick on most of your body, but up to 4 to 5 cells thick in thicker areas like the palms and soles of the feet.

Skin cells in the stratum granulosum act like specialized factories, producing tiny structures called lamellar bodies that are packed with lipids (fats) and enzymes. As these cells mature and migrate toward the skin's surface, the lamellar bodies release their lipid contents into the spaces between cells. This is the cellular “mortar” for the "bricks" (dead corneocytes filled with keratin) that create the water-repelling feature of your stratum corneum skin barrier layer.

The stratum granulosum's lipid production process is critical for optimal skin health and function. When this layer functions poorly due to genetics, damage, or aging, they lead to increased Transepidermal Water Loss (TEWL), dryness, and more potential entry points for irritants and pathogens. Poor functioning can also contribute to the development or worsening of skin conditions like eczema and psoriasis.

When the stratum granulosum is not functioning optimally, it is important to use a well-formulated gentle cleanser (harsh cleansers can cause further damage) and a moisturizer with quality natural moisturizing factor (NMF) ingredients to help restore your skin's lipid matrix. The lipids and enzymes produced in the stratum granulosum layer depend on a strong structural foundation, which is precisely what the next layer, the stratum spinosum, provides.

Stratum Spinosum: Keeping Skin Strong

The stratum spinosum is also called the "spiny layer" because of the way it looks under a microscope. The cells look spiny because they're connected by thousands of tiny molecular bridges called desmosomes. These connections work like super-strong Velcro, holding skin cells together so tightly that your skin can stretch, bend, and move without falling apart. Without this structural foundation, your skin would literally come undone every time you smiled or moved.

The cells in this layer perform multiple essential functions. They're still alive and actively producing keratin, the tough protein that gives your skin its strength and durability. As cells in the spiny layer pump out more keratin, they're essentially building the materials that will eventually form your protective outer barrier. This layer also houses important immune cells called Langerhans cells, which act like security guards, constantly patrolling for threats and alerting your immune system when they detect harmful invaders.

When your stratum spinosum layer gets damaged, whether from over-exfoliation, harsh scrubbing, or aggressive treatments, the results are noticeable and long-lasting. Your skin loses its structural integrity, becoming more sensitive, prone to irritation, and slower to heal. The damage can take weeks to fully repair because your skin has to rebuild those essential cellular connections from scratch. A damaged stratum spinosum benefits from anti-inflammatory ingredients and those that promote cell regeneration and turnover.

This strong structural foundation provided by the stratum spinosum depends on a steady supply of new cells, which is precisely what the deepest layer of your epidermis, the stratum basale provides.

Stratum Basale: The Skin Factory Where New Cells Start

The stratum basale, also called the stratum germinativum (meaning "germinating layer"), is where your skin's renewal begins. Basal cells are cuboidal-shaped (rectangular prism or a box shaped) stem cells that constantly undergo cell division to produce new keratinocytes. These plump, round cells spend the next 28 to 45 days traveling upward through your skin's layers, transforming along the way until they reach the surface as protective, flattened cells that eventually shed.

The stratum basale also contains melanocytes that create the melanin that pigments your skin and Merkel cells. Merkel cells are oval-shaped cells that work like tiny pressure sensors, detecting light touch, texture, and gentle pressure against your skin. While they make up only a small portion of the cells in your stratum basale, they play a vital role in how you experience and interact with the world around you through touch.

Each Merkel cell connects to nerve endings, instantly sending touch information to your brain so you can distinguish between a soft cotton shirt and coarse sandpaper - without even looking. The melanocytes in your skin's stratum basale are your pigment-producing cells. These cells have long, branching arms that reach out like octopus tentacles, delivering packets of melanin (pigment) to surrounding keratinocytes.

Melanin produced by the melanocytes protects your skin by absorbing and scattering harmful ultraviolet (UV) radiation from the sun. This absorption helps prevent UV rays from penetrating deeper layers of the skin, where they can damage DNA and potentially lead to skin cancer. Melanin also acts as an antioxidant, neutralizing harmful free radicals produced by UV exposure. When you get a tan, spend time in the sun, or even experience inflammation from a pimple, melanocytes produce more melanin to protect your skin from damage.

The health of your stratum basale layer determines how well your entire skin functions. When this cellular birthplace is working optimally, it produces new cells at a steady, reliable pace. Your skin looks smooth, heals efficiently, and maintains an even tone. The stratum basale layer can be damaged from excessive exposure to the sun's UV rays, chemical burns, inflammation such as dermatitis or autoimmune conditions, aesthetic procedures like aggressive peels or ablative laser treatments, or just worn out through normal aging.

Damage causes chaotic and irregular cell production that impairs skin renewal, causing excess pigment production and slow skin barrier recovery. This leads to uneven skin tone, increased sensitivity, stubborn hyperpigmentation, and health issues such as poor wound healing, When DNA is damaged, particularly from excessive UV exposure, cells can become cancerous and basal cell carcinoma can develop.

Because your stratum basale sits at the base of the epidermis, repairing it isn't as simple as applying a moisturizer. Optimal recovery requires working internally to ensure adequate nutrients and externally to facilitate healing. Just like every part of your body, your stratum basale has nutritional needs that your body must supply. Poor diet, illness, and age-related absorption are some of the reasons nutrients needed can become deficient.

Healthy skin needs lots of different nutrients, and there are books written on the subject. Zinc and vitamin C are two very important nutrients for supporting your stratum basale health. Zinc facilitates DNA synthesis, protects against oxidative stress, and regulates cell production. Vitamin C promotes collagen production, neutralizes free radicals, and helps keratinocytes transform into the specialized cells your skin needs to be healthy.

Externally, diligently using a broad-spectrum SPF 30 or higher to protect your stratum basale from excessive sun exposure is step one. To help repair the stratum basale, skincare needs to focus on restoring the barrier, stimulating cellular regeneration, and protecting against further stress. This means avoiding harsh exfoliants, retinoids, or aggressive skin treatment procedures that can cause further trauma during your skin's healing process.

Today there are skincare products with regenerative signaling ingredients like peptides and growth factors that help promote healthy basal-cell turnover and guard against oxidative stress. Professional topicals applied to micro-injuries such as microneedling, microchanneling, and fractional lasers dramatically improve penetration of regenerative ingredients to the deeper dermal layers where they can most powerfully kick-start stratum basale rejuvenation.

While the stratum basale layer serves as the birthplace for all your skin cells, it sits on an even deeper foundation that provides the structural support and nourishment these cells need to thrive, your dermis.

Dermis: Collagen, Strength, and Bounce

Your dermis is 15 to 40 times thicker than your epidermis. It houses collagen and elastin for strength and flexibility, plus blood vessels, nerve endings, sweat glands, and hair follicles. It contains all the infrastructure that keeps your skin strong, flexible, and alive.

Your dermis is where your skin structure is supported with collagen fibers that provide strength and elastin fibers that allow your skin to snap back into place after being stretched or compressed. It is where networks of blood vessels deliver oxygen and nutrients to your skin cells while carrying away waste products. Nerve endings throughout this layer provide sensory feedback, letting you feel temperature, pressure, and pain.

The roots of your hair follicles are in your dermis layer. So are your sebaceous glands that produce the natural oils that help keep your skin waterproofed and protected. The sweat glands located in the dermis help regulate your body temperature while also contributing to your skin's natural protective acid mantle. Perhaps most importantly for how your skin looks and feels, your dermis contains glycosaminoglycans (GAGs) such as hyaluronic acid. These molecules can hold up to 1,000 times their weight in water, and they are what give your skin that plump, hydrated, youthful appearance.

Fibroblasts, the primary cells in the dermis's connective tissue, are essential for wound healing, migrating to injury sites to produce collagen, elastin, and extracellular matrix (ECM) components that rebuild tissue structure. They also secrete growth factors, cytokines, and signaling molecules to orchestrate tissue repair and regeneration.

Your dermis can get damaged from UV exposure, inflammation, or injury. It also deteriorates with age, naturally producing less collagen, elastin, and hyaluronic acid. Whether from damage or natural aging, signs of deteriorated dermis health are wrinkles, sagging skin, and skin volume loss.

Supporting and repairing the dermis requires a focus on stimulating fibroblast activity, enhancing collagen and elastin synthesis and protecting the skin from further breakdown. The skincare ingredients that stimulate collagen and elastin production include retinoids, vitamin C, niacinamide, growth factors, and peptides, the same as for your stratum basale layer.

Applying skincare products with peptides, especially copper peptides, at the same time as retinol, vitamin C, and alpha hydroxy acids like glycolic acid reduce effectiveness and can cause irritation. If you're using both, use retinol and other acids at night and copper peptides in the morning, or apply them on separate days. How to pair actives, avoid clashes, and ingredient application timing is covered in the *Smart Skincare Parings for Better Results* section a little further ahead.

Now that we've explored the dermal structure, let's look at the powerful messengers that orchestrate repair across all layers, growth factors and exosomes.

Hypodermis: Your Skin's Protective Padding

Below your dermis lies the deepest layer of your skin structure, your hypodermis - also known as the subcutaneous layer. This layer isn't just passive padding; it's a metabolically active foundation that influences everything from your skin's appearance to your overall health. The hypodermis is primarily composed of adipose tissue (fat cells) organized into lobules separated by fibrous connective tissue.

These fat cells do far more than just store energy; they act as your body's internal cushion, regulate temperature, and function as a hormone-producing organ. They secrete adipokines, signaling molecules that can influence inflammation, immune responses, and skin healing. When your hypodermis is healthy, it produces anti-inflammatory signals that support skin renewal. When compromised by chronic inflammation, poor diet, or metabolic dysfunction, the hypodermis releases pro-inflammatory signals that accelerate skin aging and worsen conditions like acne or eczema.

The hypodermis also determines much of what we perceive as youthful skin structure. The distribution and volume of fat in this layer create contours that define a youthful appearance. It's what creates full cheeks, a smooth jawline, plump hands, and the absence of deep folds or hollows. As you age, the fat in this layer doesn't just shrink; it redistributes in predictable patterns. You lose volume in areas like the temples, cheeks, and hands while gaining it in areas like the lower face and neck.

This redistribution, combined with overall volume loss, creates structural changes that no topical product can fully address. Several factors accelerate this deterioration: chronic inflammation from poor diet or stress, UV exposure that triggers inflammatory cascades that affect deeper layers, and smoking that reduces blood flow while increasing enzymes that break down supporting connective tissue.

While traditional skincare products don't penetrate to the hypodermis, you can still support this layer indirectly. Maintaining a strong skin barrier helps prevent inflammatory signals from penetrating deeper, while broad-spectrum sunscreen protects against UV-induced inflammation affecting all skin layers. The most effective support comes from lifestyle factors, including a diet rich in anti-inflammatory foods, regular exercise to improve circulation, and quality sleep for repair and regeneration.

For concerns directly related to hypodermis changes like volume loss or deep folds, topical products have limitations. The newest professional regenerative products help skin function better at a cellular level; however, they need a controlled injury like microneedling to reach the deeper layers of the skin where they are most effective. Treatments like High-Intensity Focused Ultrasound (HIFU) and monopolar radiofrequency can reduce the fat that has been redistributed to the lower face and neck. Deeper facial folds and wrinkles can be visibly minimized with professional interventions like dermal fillers or fat transfers to increase volume.

Understanding your hypodermis helps explain why not all skin concerns can be addressed topically. When volume, structure, or deep support is compromised, targeted regenerative treatments along with strategic skincare products are often required to restore skin health from the foundation up.

Cell Talk: Messengers That Guide Skin Repair

Growth factors and exosomes are molecular messengers that coordinate repair and regeneration. They are produced all throughout the epidermis by keratinocytes, melanocytes, Langerhans cells, and basal cells, in the dermis by fibroblasts, endothelial cells, macrophages, and mast cells, and in the hypodermis by mesenchymal stem cells. You don't need to remember all that. You just need to remember that they are naturally produced by many different types of cells that are distributed across all your skin layers and throughout your body.

Growth factors are special types of protein made naturally by your body's cells. Think of them as a messenger or "instruction manual" that tells skin cells when to repair themselves, grow, and make more of important building blocks like collagen and elastin. In skin physiology, several different human growth factors play specific roles.

- **Epidermal Growth Factor (EGF):** Stimulates the production of new skin cells and helps repair the outer layer of skin, aiding in healing and regeneration. EGF is widely used in skincare serums and creams for its ability to stimulate skin cell proliferation and wound healing.

- **Fibroblast Growth Factor (FGF):** Supports the production of collagen, elastin, and other building blocks in the skin, foundational for firmness and elasticity. It is used in some skincare products to help promote collagen and elastin production.

- **Transforming Growth Factor Beta (TGF-β):** Promotes collagen formation and helps maintain the skin's structure. It is also involved in wound healing and scar regulation. Some skincare formulations use it for its collagen production and scar regulation properties, most often in products that contain human fibroblast conditioned media:

- **Vascular Endothelial Growth Factor (VEGF)**: Encourages the formation of new blood vessels, which support skin nourishment and repair. VEGF's use in skincare products is controversial due to its elevated levels in melanoma patients, which raises concerns about its cancer risks. Its inclusion is limited to specialized medical products, and dermatologists advise caution.

- **Keratinocyte Growth Factor (KGF)**: Stimulates the growth of keratinocytes, the predominant cells in the outer layer of the skin, which helps in surface repair and resilience. It is used in some skincare products for its specific action that promotes skin surface repair and hair health.

- **Platelet-Derived Growth Factor (PDGF)**: Regulates cell growth and division, particularly important during wound healing and tissue maintenance. PDGF is used in some professional skincare and prescription wound-healing products.

- **Insulin-like Growth Factor 1 (IGF-1)**: Stimulates skin cell growth and proliferation. It is rarely used in skincare due to strict regulatory restrictions and potential safety concerns, including theoretical risks of promoting cancer cell growth or exacerbating inflammatory conditions like acne, primarily associated with systemic exposure.

- **Basic Fibroblast Growth Factor (bFGF)**: Helps form new blood vessels and supports skin cell regeneration.

- **Granulocyte-Monocyte Colony-Stimulating Factor (GM-CSF)**: Increases the number of protective white blood cells and strengthens the skin barrier. This naturally occurring growth factor is not used in skincare products due to its proinflammatory nature and potential risks, such as worsening inflammatory skin conditions or promoting tumor growth.

- **Interleukins (IL-6, IL-7, IL-8)**: A group of proteins that manage inflammation and modulate immune responses in the skin. These potent inflammatory cytokines - part of the body's natural healing response - are not used in skincare products.

Each of these growth factors sends specific instructions to the skin, working together to keep it youthful, firm, and able to heal from damage or irritation. They are naturally made by your body, and today they are increasingly available in advanced skincare products to support visible repair, texture, and resilience.

Exosomes Explained

In biology, vesicles are tiny, bubble-like structures that cells naturally create to carry messages and materials. One special type of vesicle is called an exosome. Exosomes are ultra-small vesicles - think of them as tiny envelopes that cells use to send messengers like growth factors to other cells. When one cell releases exosomes containing growth factors, another cell can absorb them and "read" the contents, kind of like receiving a text message.

As you age, your cells produce and fewer growth factors and exosomes that transport them. Fewer growth factors mean decreased cellular communication, which is one reason why skin renewal slows from a youthful 28-day cycle to 45-50 days in mature skin. The dermal fibroblasts, which produce collagen, elastic fibers, and other extracellular matrix (ECM) components become less responsive to repair signals causing collagen production to diminish.

In skincare and regenerative medicine, exosomes are exciting because they can deliver growth factors that help other cells heal, produce more collagen, or reduce inflammation to help your skin repair itself from the inside out. When applied topically in high concentrations after a controlled injury like microneedling, growth factors and exosomes can penetrate deeply, initiating a communication cascade that extends all the way down into your dermis.

It only takes a tiny amount to trigger a chain reaction that ultimately stimulates dermal fibroblasts to produce new collagen and elastin. This process helps restore more youthful cellular behavior, meaning faster turnover in your epidermis, increased collagen synthesis in your dermis, and improved barrier function throughout your skin structure.

Not All Growth Factors Are Equal: Why Source Counts

The origin of growth factors determines how they interact with human skin cells. Plant growth factors, like those found in some skincare products, are molecules that plants use to regulate their own growth and development. Human cells have receptors tailored to human growth factors (e.g., epidermal growth factor), which are part of our natural cellular communication system. Without compatible receptors, plant growth factors cannot directly stimulate human cell growth or repair in the same way.

Skincare ingredients derived from human adipose tissue (fat cells), bone marrow, or umbilical cord stem cells are rich in growth factors like EGF and TGF-β that potently enhance cellular repair, including stimulating fibroblasts to produce collagen and elastin. Plant stem cell extracts, derived from sources like apple or Centella asiatica, mildly mimic these effects through peptides and antioxidants that support collagen and elastin synthesis.

The most effective formulations contain multiple types of growth factors working together, rather than isolated single factors. Your skin's repair processes require complex orchestration, like a symphony rather than a solo performance. Products that contain comprehensive growth factor profiles can support the intricate cellular communication needed for optimal skin function across all layers.

While topical growth factors and exosomes provide measurable improvements in skin texture, fine lines, and overall appearance, their penetration depth is naturally limited. Their primary effects occur in your epidermis and upper dermis, where they can restore healthier cellular turnover patterns and stimulate collagen production.

For deeper effects that reach your hypodermis, growth factors are most effective when combined with delivery enhancement methods like microneedling, microchanneling, or professional treatments that temporarily compromise your skin barrier to allow deeper penetration.

Understanding growth factors and exosomes helps explain why some skincare products can produce dramatic improvements in skin appearance. They're not just adding ingredients to your skin's surface but reprogramming your cells to behave more like younger, healthier skin. This represents a fundamental shift from traditional skincare that simply treats symptoms to advanced formulations that address the underlying cellular communication breakdowns that drive skin aging.

Odds & Ends That Matter

Every product you apply to your skin makes a difference - sometimes for the better, sometimes not. These additional fundamental concepts will help you make more confident choices.

The Acid Mantle: Your Skin's Natural Defense System

Your skin maintains a slightly acidic surface pH of approximately 4.5 to 5.5. This isn't accidental. It's essential for optimal function. This acidic environment, called the acid mantle, serves multiple purposes. It discourages harmful bacteria while supporting beneficial microbes. It optimizes enzyme activity for lipid production. It helps maintain the structural integrity of your barrier.

When you use high-pH products (above 7), you temporarily alkalize your skin surface. This disruption can last hours, during which time your skin becomes vulnerable to bacterial invasion and moisture loss. Understanding pH helps you choose products that work with your skin's chemistry rather than against it.

Your Skin's Microbiome: Why Good Bacteria Matter

Your skin hosts millions of beneficial microorganisms that form a protective ecosystem. These bacteria produce antimicrobial compounds, maintain optimal pH, and train your immune system.

Disrupting this balance with harsh antimicrobials or high alcohol concentrations can create dysbiosis. This is an imbalance that can cause symptoms like skin sensitivity, breakouts, or chronic inflammation. Rather than sterilizing your skin's surface, effective skincare nurtures beneficial bacteria while maintaining cleanliness.

What Gets In? The 500-Dalton Rule That Decides Which Ingredients Work

Not all ingredients can penetrate your skin barrier, and that's intentional. The "500 Dalton Rule" suggests that molecules larger than 500 Daltons, about the mass of three glucose molecules or a small peptide with 4–5 amino acids, generally cannot penetrate intact skin.

This explains why popular ingredients like collagen (300,000 Daltons) act only on the surface, while smaller molecules like retinol (286 Daltons) can reach deeper layers where they create cellular changes.

Understanding molecular size helps set realistic expectations. Surface-acting ingredients provide immediate benefits through hydration and protection. Smaller molecules deliver long-term changes through cellular communication.

What Your Skin Is Trying to Tell You

Your skin is as unique as your fingerprint, and unlocking its potential starts with understanding its specific needs. When you understand your skin's unique profile, you can choose formulations that work with, rather than against, your individual biology.

Dermatologists commonly evaluate skin through a structured, three-pillar framework that anyone can use. The first pillar is skin barrier health, which reflects how well the skin's protective layer maintains hydration and defends against irritants. The second is skin type: the genetically influenced baseline - oily, dry, sensitive, or combination - that determines how your skin naturally functions.

The third pillar is primary concerns, the visible conditions or issues that you want to improve, such as acne, redness, pigmentation, or premature aging. Together, these three areas create a clinically grounded yet practical roadmap for understanding your skin and making smarter choices about how to care for it.

Step One - Assess Your Skin Barrier Health

The state of your skin barrier determines both what your skin can tolerate and what it desperately needs. This assessment should always come first; it influences every other ingredient decision.

Your skin barrier consists of lipid layers that seal moisture in and irritants out. When functioning properly, it maintains optimal hydration levels, regulates temperature, and protects against environmental damage. When compromised, it creates a cascade of conditions that affect how every product performs.

Physical signs of a compromised skin barrier include tightness, flaking, or persistent dryness that doesn't improve with moisturizers. You might notice increased sensitivity to previously tolerated products, or stinging or burning sensations from ingredients that never bothered you before. Frequent irritation from environmental factors like wind, air conditioning, or heating systems also signal barrier dysfunction.

From a functional standpoint, compromised skin barrier indicators include slow healing from breakouts or minor injuries. You may notice that skincare products seem to "sit on top" of your skin rather than being absorbed, or that moisturizers provide only temporary relief before dryness returns. When skin barrier health is compromised, its repair needs to be a priority.

Step Two - Determine Your Underlying Skin Type (*phenotype*)

Your skin type determines how products feel, absorb, and perform on your skin, influencing both comfort and effectiveness. While skin type can shift over time due to age, hormones, climate, or other factors, understanding your current type guides immediate product selection decisions.

This is a simple and reliable "bare face" method you can do at home:

1. Cleanse Your Face
Wash your face with a gentle, non-stripping cleanser to remove dirt, oil, and makeup without disrupting your natural skin barrier.

2. Don't Apply Any Products
Gently pat your skin dry, do not apply moisturizers, serums, or toners. Leave your skin completely bare.

3. Wait and Observe (60–90 minutes)
Let your skin rest for about an hour. During this time, avoid touching your face.

4. Examine Your Skin
Stand in natural light or use a mirror in a well-lit area and assess:

> **Oily skin**: Skin appears shiny, especially in the T-zone (forehead, nose, chin). Pores may look enlarged. You may feel a greasy film or notice that blotting tissue picked up oil.
>
> **Dry skin**: Skin feels tight, rough, or flaky. You may see dullness or red patches. Fine lines may appear more prominent due to dehydration.
>
> **Combination Skin**: Oily in the T-zone, but dry or normal on the cheeks or jawline. This is the most common skin type.
>
> **Normal skin**: Feels balanced, not dry, not oily. Smooth texture, small pores, no major sensitivities or flaky areas.

It's important to remember that your skin type is not a "one and done". Most people need to make seasonal adjustments at a minimum, using lighter formulations in the summer and richer textures in the winter.

Step Three - Identify Primary Concern(s)
Your primary skin concerns determine which active ingredients to prioritize and how to sequence them in your routine. While it's tempting to address multiple concerns simultaneously, focusing on one or two top priority issues typically produces better results than scattered approaches. Step 3 is picking your number one, and maybe a number two concern.

Integration Strategy - Putting It All Together
Understanding each factor on its own is useful, but the real clarity comes from integrating all three pillars into an ingredient-focused strategy. This means evaluating barrier health, skin type, and concerns together, and then choosing ingredients that respect those relationships.

Always start with barrier health, since it sets the boundaries for ingredient tolerance and determines how quickly you can move toward your goals. When the barrier is compromised, even ingredients that are ideal for your skin type or concern, such as retinoids, acids, or vitamin C, can worsen sensitivity or inflammation.

Next, take your skin type into account because it guides formulation choice as much as ingredient choice. or instance, oily skin may respond best to lightweight, oil-free formulations that deliver actives like niacinamide or salicylic acid without added weight, while dry skin benefits from richer textures paired with barrier-replenishing ingredients like ceramides and hyaluronic acid.

Finally, narrow your focus of the skincare ingredients you'll be looking for to those to one or two primary concerns. If acne is your priority, you might choose salicylic acid or benzoyl peroxide; for pigmentation choose niacinamide or azelaic acid; for fine lines choose retinoids or peptides. Your concerns define what you're targeting, but barrier health and skin type dictate how you incorporate those actives safely and effectively.

When used together, these three pillars are the first step in transforming ingredient selection from guesswork into a clear, clinically grounded process- one that balances ambition with tolerance and delivers sustainable, long-term results.

Building Your Perfect Skincare Routine

Skincare is not one-size-fits-all; effective routines are tailored to individual needs that change over time, blending products from various categories - retail, professional, and/or DIY - based on informed choices and skin observation.

A strategic hybrid approach prioritizes spending on products with prolonged skin contact, such as serums or occlusive moisturizers like shea or mango butter, which often outperform costly alternatives for specific concerns.

Daily cleansers, with minimal skin contact time, typically don't warrant premium pricing, while advanced professional topicals with high concentrations of targeted ingredients for addressing stubborn conditions like acne, pigmentation, eczema, or psoriasis can be well worth the extra spend. Consulting a skincare professional can help those with persistent issues integrate targeted professional products with compatible retail or DIY options, optimizing results within budget constraints. There's more about that in the upcoming *When to Go Professional* section.

Building an effective routine goes beyond selecting quality products; it requires understanding strategic application - layering products by molecular logic, choosing synergistic combinations, and avoiding counterproductive pairings. The following sections outline key principles for applying products, from proper layering techniques to safely introducing new actives, monitoring skin responses, and adapting routines as your skin evolves, empowering you to achieve lasting skin health.

Smart Skincare Pairings for Better Results

Skincare ingredients work best when thoughtfully combined to enhance their effects or avoid irritation. The right pairings can boost results for concerns like aging, acne, or dryness, while the wrong ones may cause redness, sensitivity, or reduced effectiveness. By understanding which ingredients work well together and which to keep apart, you can build a routine that maximizes benefits and keeps your skin healthy and happy.

Layering & Timing

To get the most out of your skincare products, apply them in the right order and give them time to work. After cleansing, start with lightweight products like toners or essences to prep your skin.

Next, use serums with ingredients like niacinamide, peptides, or panthenol, which need a neutral skin environment to perform best. Then, apply hydrating serums like hyaluronic acid to lock in moisture.

If you're using stronger actives like vitamin C or exfoliating acids (glycolic or salicylic), wait 15–30 minutes after the first serums to avoid irritation, such as redness from mixing niacinamide with acids.

Finish with thicker products like creams, lotions, or oils to seal everything in and protect your skin's moisture barrier. This simple step-by-step approach helps each product work better, leaving your skin healthier and glowing.

Great Combinations

These pairings work together like a team, boosting each other's benefits, reducing irritation, or improving how well products absorb. They're perfect for tackling aging, acne, pigmentation, or dryness more effectively than using ingredients alone.

- **Exfoliating Acids (AHAs + BHAs)**: These exfoliants work at different skin depths to comprehensively clear dead skin and debris; AHAs (glycolic acid, lactic acid, citric acid, and mandelic acid) work on the surface while BHAs (salicylic acid, willow bark extract, betaine salicylate) can actually penetrate into pores.
- **Bakuchiol + Exfoliating Acids**: Gentle exfoliation from acids helps bakuchiol (a retinol alternative) absorb better, improving fine lines and texture without the irritation of retinol.
- **Centella Asiatica + Peptides**: Centella soothes and heals, while peptides boost skin repair, speeding up recovery for irritated or damaged skin.
- **Ceramides + Cholesterol + Fatty Acids**: The ideal ratio (3:1:1) mimics skin's natural barrier composition, enhancing barrier repair more effectively than any component alone.
- **Glycolic Acid + Salicylic Acid**: When formulated at appropriate pH levels, this combination provides both surface renewal and pore-clearing benefits for combination skin types.
- **Green Tea Extract + Resveratrol**: These antioxidants work through different pathways to provide comprehensive protection against various types of free radical damage.
- **Hyaluronic Acid + Moisturizers**: Hyaluronic acid pulls in moisture, and a thick moisturizer (like shea butter) locks it in, keeping skin hydrated and plump, especially in dry climates.

- **Niacinamide + Zinc:** Together, they control oil, calm redness, and support the skin barrier, making them a go-to for acne-prone skin.
- **Peptides + Antioxidants:** Antioxidants protect peptides from breaking down while peptides help rebuild what free radicals damage, creating a comprehensive anti-aging approach.
- **Retinol + Niacinamide:** Niacinamide reduces the irritation potential of retinol while both help improve skin texture and tone; niacinamide also supports the skin barrier function.
- **Retinol + Peptides:** While retinol stimulates collagen production, peptides help organize new collagen fibers properly for more effective skin remodeling.
- **Squalane + Humectants:** Humectants draw water into the skin while squalane prevents that moisture from evaporating, creating lasting hydration.
- **Vitamin C + Alpha Arbutin:** Both target hyperpigmentation through different mechanisms; vitamin C suppresses melanin production while alpha arbutin blocks tyrosinase activity.
- **Vitamin C + Sunscreen:** Morning application of vitamin C under sunscreen boosts photoprotection by neutralizing free radicals not blocked by SPF, providing up to 8x additional protection.
- **Vitamin C + Vitamin E + Ferulic Acid:** This antioxidant trio enhances stability and effectiveness (vitamin E improves vitamin C stability by up to 4x, while ferulic acid further stabilizes both and boosts UV protection).
- **Vitamin E + Ferulic Acid:** This antioxidant duo teams up to protect your skin from damage caused by sun and pollution, which can lead to wrinkles and dullness.

Combinations to Avoid or Use Carefully

Some ingredients don't play well together, causing irritation that reduces effectiveness and can harm your skin barrier. Keep these apart or use them at different times to avoid problems.

- **Exfoliating Acids + Vitamin C:** Both work in acidic environments but may increase sensitivity when used together. If using both, select a properly formulated product that combines them rather than layering separate products.
- **Exfoliating Acids + Clay Masks:** Following acidic exfoliants with drying clay masks can severely dehydrate skin and trigger inflammation or rebound oil production. Space these treatments by at least 24 hours.
- **Benzoyl Peroxide + Retinol:** Benzoyl peroxide can oxidize and deactivate retinol molecules, rendering both ingredients less effective. Separate by using one in morning, one at night, or on alternate days.
- **Acids + Retinoids:** Can cause excess irritation, barrier damage, and increased sensitivity; better used on alternate days or after building tolerance. If combined, start with a just once weekly application.
- **Combined Potential Irritants:** Products containing alcohols, fragrance, essential oils, or preservatives like methylisothiazolinone may be tolerable individually but sensitizing when combined, creating a cumulative irritation effect.

- **Fragrant Essential Oils + Sensitive Skin Ingredients:** Ingredients meant to calm sensitivity like ceramides can have their benefits undermined by potentially sensitizing fragrant oils like citrus or mint oils.
- **Multiple Strong Actives:** Overloading your skincare regimen with ingredients like retinol, vitamin C, and acids at once can overwhelm skin, causing redness or flaking. Stick to one or two actives at a time.
- **Oils Before Serums:** Applying oils before water-based serums blocks their absorption. Always use oils after serums or as the last step.
- **Peptides + Acids:** Acidic products can break down peptide bonds, reducing efficacy. If using both, apply acids first, wait 30 minutes for pH to normalize, then apply peptide products.
- **Physical Exfoliation + Chemical Exfoliation:** Combining these exfoliation methods in the same session can damage the skin barrier. Space manual scrubs and acid treatments by at least 72 hours.
- **Retinoids + Benzoyl Peroxide + Salicylic Acid:** This "triple therapy" approach is extremely harsh, damaging the moisture barrier and potentially worsening acne through irritation inflammation. Choose maximum two of these for any routine.
- **Vitamin A Derivatives Together:** Combining multiple forms of vitamin A (retinol, retinal, retinoic acid, etc.) significantly increases irritation risk without proportional benefits. Choose one form rather than layering multiples.
- **Vitamin C + Metal-Based Products:** Vitamin C can interact with metal ions in skincare (copper peptides, zinc oxide), causing oxidation or reduced efficacy. Wait at least 10 minutes between applications.
- **Vitamin C + High-Strength Niacinamide:** In some cases, combining high doses (e.g., 10%+ vitamin C with 5%+ niacinamide) can cause temporary redness. Use them at different times or in products designed to work together.
- **Vitamin C (L-Ascorbic Acid) + Retinol:** These work at different pH levels, which can reduce their benefits or cause irritation. Use vitamin C in the morning and retinol at night.

Keep It Simple for Better Skin

More isn't always better in skincare. Choosing a few ingredients that work well together is often smarter than piling on multiple products. Pay attention to how your skin feels and introduce new products slowly to avoid irritation. With the right combinations, you'll see better results and keep your skin healthy and glowing.

Patch Testing & Tracking Results

Introducing new skincare products can be exciting, but it's important to test them carefully and track changes to ensure they work for your skin. Patch testing helps you avoid irritation and allergic reactions; tracking progress lets you see how your skin improves over time. By following a simple, step-by-step approach, you can build a routine that's safe and effective, giving you healthier, happier skin.

Patch Testing: A Safe Start

Patch testing checks if a product might irritate or cause an allergic reaction, is especially important for those with sensitive skin. Patch testing is more than just a quick dab - here's how to do it right:

1. **Pick a test spot:** Start with the inner forearm or behind the ear for an initial check. For a closer match to your face, try the jawline or side of the neck.
2. **Apply daily:** Put a small amount of the product on the same spot for 3–7 days, as some reactions take time to show up.
3. **Watch for issues:** Look for redness, itching, burning, or bumps. A little tingling from active ingredients like acids or vitamin C is often normal if it fades quickly.
4. **Test on your face:** If the spot test is clear, apply the product to a small area of your face (like one cheek) for another 3–7 days before using it all over.

If you have very sensitive skin or a history of reactions, test products with single active ingredients to make it easier to pinpoint any triggers. This careful approach helps you avoid tossing out a whole product if only one ingredient causes trouble.

Tracking Your Skin's Progress

Skincare changes can be slow, so tracking helps you notice improvements and stay on the right path. Here's how to keep tabs on your skin:

- **Take starting photos:** Snap clear pictures of your skin before using new products. Use the same lighting, angle, and expression each time, with close-ups of problem areas like acne or dark spots.
- **Check progress regularly:** Look at your skin every 2–4 weeks instead of daily to spot real changes, like smoother texture or less redness.
- **Notice how skin feels:** Track changes in hydration, smoothness, or comfort, as these often improve before you see visible results.
- **Keep a skincare journal:** Write down when you start new products, any reactions, and how your skin feels over time. Note things like stress, weather, or hormonal changes that might affect your skin.

How Long Until You See Results?

Different ingredients take time to work, so patience is key. Here's a simple guide to when you might see changes with common skincare ingredients:

- **Hydrators** (e.g., glycerin, hyaluronic acid): Immediate to 1 week for softer, plumper skin.
- **Exfoliants** (e.g., glycolic acid, salicylic acid): 2–6 weeks for smoother skin and clearer pores; dark spots or acne take longer.
- **Antioxidants** (e.g., vitamin C, ferulic acid): 4–12 weeks for brighter skin and fewer dark spots.
- **Brighteners** (e.g., niacinamide, alpha arbutin): 6–12 weeks for more even skin tone.
- **Acne fighters** (e.g., benzoyl peroxide, azelaic acid): 4–8 weeks to reduce breakouts; early "purging" may happen.
- **Barrier helpers** (e.g., ceramides, squalane): 1–4 weeks for less sensitivity and smoother skin.
- **Anti-aging** ingredients (e.g., retinol, peptides): 8–16 weeks for fewer fine lines and firmer skin.
- **Eye creams** (e.g., caffeine, hyaluronic acid): 4–8 weeks for less puffiness; dark circles or wrinkles take longer.

- **Sunscreen** (e.g., zinc oxide): Instant protection from sun damage; prevents aging over time.

Skin takes about a month to renew itself (longer as you age), so most ingredients need at least one cycle to show results. Stick with your routine, and you'll see your skin improve with time and consistency.

Purging vs. Reactions: Know the Difference

When you start using active skincare ingredients like retinoids or exfoliating acids, your skin might react in ways that seem concerning. Knowing the difference between normal "purging" (temporary breakouts as skin adjusts) and harmful reactions (irritation or allergies) helps you decide whether to keep using a product or stop. This guide makes it easy to spot the signs and adjust your routine for healthy, happy skin.

Signs of Normal Purging

Purging happens when ingredients speed up skin cell turnover, bringing hidden clogged pores to the surface. It's temporary and usually improves with time. Look for:

- **Breakouts in usual spots**: Small pimples or bumps appear where you typically get acne.
- **Quick resolution**: These breakouts clear up relatively fast, often within days.
- **Improvement over time**: Skin looks clearer and smoother after 4–6 weeks.
- **Common with actives**: Happens with ingredients like retinoids (e.g., retinol), exfoliating acids (e.g., glycolic or salicylic acid), or other cell-turnover boosters.

Signs of a Harmful Reaction

A reaction means the product is causing irritation or sensitivity. Stop using the product if you notice:

- **New breakout areas**: Redness, bumps, or rashes in places that you don't usually have any issues.
- **Ongoing discomfort**: Persistent redness, itching, burning, or tight, dry skin.
- **Worsening symptoms**: Problems get worse the longer you use the product.
- **Uncomfortable sensations**: Prolonged stinging, flaking, or a tight feeling that doesn't improve.

What to Do

- **For purging**: Keep going but ease up. Consider using the product every other day instead of using it daily or only applying it over a light moisturizer to reduce intensity while your skin adjusts.

- **For reactions**: Stop using the product right away. If irritation persists, consult a dermatologist for advice. Always patch-test new products (apply to a small area for 3–7 days) to catch reactions early.

By recognizing these signs, you can stick with products that work and avoid those that harm, building a routine that leaves your skin clearer and healthier.

Adjusting Your Skincare for Changing Needs

Your skin needs shift with seasons, age, hormones, and lifestyle, so a smart skincare routine evolves to keep your skin healthy and balanced. By tweaking your products based on weather, indoor environments, or life changes, you can address challenges like dryness, oiliness, or sensitivity. Here's a simple guide to adapting your year-round routine for glowing, comfortable skin.

Summer Adjustments

Hot, humid weather and stronger UV rays call for lighter products and extra protection:

- **Use lightweight moisturizers**: Choose hydrating ingredients like hyaluronic acid or glycerin instead of thick creams to keep skin refreshed without feeling heavy.
- **Boost sun protection**: Add antioxidants like vitamin C or ferulic acid under sunscreen to fight sun damage and prevent dark spots or wrinkles.
- **Ease up on strong actives:** If retinol or exfoliating acids (like glycolic or salicylic) feel irritating, use them less often (e.g., every other night).
- **Gently exfoliate**: Use mild exfoliants to control extra oil and prevent clogged pores without over-drying your skin.

Winter Adjustments

Cold, dry air can make skin dry, flaky, and feel tight, so focus on locking in moisture:

- **Switch to richer moisturizers**: Use thicker creams with ceramides, shea butter, or oils to protect and repair your skin's barrier.
- **Exfoliate less**: Cut back on exfoliating acids to avoid irritation, especially if skin feels sensitive or dry.
- **Layer hydration**: Apply hydrators like hyaluronic acid or glycerin under a heavy moisturizer to trap moisture.
- **Choose gentle cleansers**: Swap foaming cleansers for creamy or oil-based ones to clean without stripping natural oils.
- **Humidify Parched Air**: The ideal indoor humidity level for healthy skin is typically between 30% and 50%. Counteracting the drying effects of both cold outdoor air and indoor heating systems helps your skin remain hydrated, strengthens its protective barrier, and soothes irritated skin.

Transitional Seasons: Spring and Fall

Spring and fall often affect skin unpredictably - adjust your skincare regimen gradually.

- **Ease into changes**: Slowly switch between light and rich products as the weather shifts to avoid shocking your skin.
- **Support your skin barrier**: Use calming ingredients like centella asiatica or niacinamide to soothe sensitivity during season changes.
- **Introduce actives carefully**: Add or reduce strong ingredients (like retinol or acids) step-by-step to prevent irritation.

Keep Your Skincare Flexible for Life's Changes

Your skin changes with the seasons, but it's also affected by indoor environments and life stages like aging, stress, or hormonal shifts. Dry air from heating or air conditioning can dehydrate your skin year-round, so keep using hydrating products like hyaluronic acid even in milder weather. Life changes like adolescence, pregnancy, or menopause call for tweaks to your routine to keep your skin healthy and glowing. By regularly checking how your skin feels and looks, you can adjust your products to match its current needs, ensuring long-lasting results.

Skincare for Different Life Stages

Your skincare needs evolve as you age, so adapt your routine to stay ahead:

- **Teens to Early 20s**: Focus on controlling oil and preventing breakouts with gentle ingredients like niacinamide or low-dose salicylic acid. Use a mild, non-drying cleanser and start with basics like sunscreen and antioxidants (e.g., vitamin C) to protect your skin.

- **Late 20s to 30s**: Begin anti-aging prevention with ingredients like peptides or vitamin C. Add gentle exfoliation, even by just adding a half teaspoon of table sugar to your gentle cleanser once a week or so, to help keep skin smooth as cell turnover slows. As you approach your 40s, adding a low-strength retinol can help boost renewal without irritation.

- **During Your 40s**: Add peptides to support skin firmness and elasticity, helping to reduce early signs of aging like fine lines. As you approach your late 40s, consider growth factors to boost cell repair and collagen production, preparing your skin for the next decade. Use brighteners like niacinamide for age spots and gradually increase actives like retinol if your skin tolerates them. Prioritize hydration with richer moisturizers like shea butter or ceramides as skin starts to get drier.

- **During Your 50s and Beyond**: As skin becomes significantly drier, focus on deep hydration and barrier repair with thick creams containing ceramides or oils. Use firming ingredients like peptides to maintain elasticity and brighteners like alpha arbutin for age spots. Incorporate advanced biotechnology ingredients like exosomes and stem cell growth factors to promote deep skin repair, improve texture, and reduce wrinkles. These advanced actives support skin renewal at a cellular level, often working best with professional treatments for enhanced results.

Life's Hormonal Changes

Hormones can change your skin's behavior, so adjust proactively:

- **Menstrual Cycles**: Increase exfoliation (e.g., salicylic acid) before breakout-prone days or add extra hydration when skin feels dry during your cycle.
- **Pregnancy**: Skip ingredients like retinol or high-dose salicylic acid, which may not be safe. Focus on gentle hydrators (e.g., hyaluronic acid) and soothing ingredients (e.g., centella asiatica).

- **Menopause**: Boost hydration and barrier support with ceramides, emollients like shea butter, and/or oils like jojoba. Use firming ingredients like peptides to help address dryness and sagging from hormonal changes.
- **Male Testosterone Changes**: Testosterone levels, which peak in young adulthood and gradually decline after age 30, affect male skin significantly. In teens and 20s, high testosterone can increase oil production, leading to acne and shininess; use oil-controlling ingredients like niacinamide or low-dose salicylic acid with a gentle cleanser. In the 30s and 40s, declining testosterone may reduce oiliness but increase dryness or sensitivity. Switch to hydrating moisturizers like hyaluronic acid or ceramides. In the 50s and beyond, lower testosterone can thin skin and slow repair. Add peptides and growth factors to support firmness and recovery.

Smart Ways to Evolve Your Routine

A great skincare routine grows with you, with small, thoughtful changes to avoid overwhelming your skin. Here's how to adapt effectively:

- **Make one change at a time**: Add or remove one product (e.g., a new serum) and use it for 2–4 weeks to see how your skin responds before trying another change.
- **Keep your go-to products**: Stick with reliable favorites (like a trusted moisturizer or sunscreen) as the core of your routine while testing new additions.
- **Rotate strong ingredients**: Alternate powerful actives to prevent irritation or stalled results. For example, use vitamin C in the morning and retinol at night, or exfoliating acids a few times a week instead of daily.
- **Check in regularly**: Every 3–6 months, review your skin's needs. If dryness, oiliness, or sensitivity changes, swap products to match.
- **Start simple**: Begin with the basics - cleanser, moisturizer, sunscreen - then gradually add actives like retinol or acids as you learn what works for your skin.

Trust Your Skin's Feedback

Skincare is both science and personal experience. Even if an ingredient has great research, it might not suit your skin, while a less-hyped product could be a game-changer for you. The goal is to find a few key products that work well together, delivering results without irritation. This tailored approach saves time and money while keeping your skin healthy and radiant.

MATCHING INGREDIENTS TO CONCERNS

Skin concerns are very different, and so are the ingredients that are best to treat them. While marketing often promotes so-called miracle products with universal benefits, effective skincare requires a more thoughtful, targeted approach. Each condition has its own biological drivers, and each demands a unique ingredient strategy to address those root causes without compromising overall skin health.

- Regardless of skin type or concern, every ingredient strategy must start with **skin barrier health**. By ensuring the skin barrier is supported, skincare ingredients can do their job safely and effectively.

- An **acne**-focused skincare regimen requires a multi-pronged approach targeting the various factors that contribute to breakouts: excess oil production, abnormal cell shedding, bacterial proliferation, and inflammation.

- An **anti-aging** focused regimen also requires a multi-pronged approach for stimulating collagen production, providing antioxidant protection, improving skin texture and tone, and optimizing skin barrier health.

- **Hyperpigmentation** requires patience and consistency because melanin reduction occurs slowly and unevenly. The approach needs to combine melanin inhibition with accelerated cell turnover and strict sun protection.

- **Sensitive skin**, including **rosacea**, requires a completely different approach that prioritizes gentleness and simplicity over aggressive correction. The goal is to maintain skin comfort while gradually building tolerance for more active ingredients.

- **Eczema** (atopic dermatitis) requires a fundamentally different approach than most other skin concerns, prioritizing barrier restoration and inflammation control over a regimen centered on active treatments. The chronic inflammatory nature of eczema means that aggressive correction can often worsen the condition, making gentle, supportive care the most effective strategy.

- **Psoriasis** requires targeted treatment that addresses accelerated cell turnover, thickened scale buildup, and underlying inflammation characteristic of this autoimmune condition. Unlike eczema, psoriasis often benefits from more active ingredients that help normalize cell production and remove accumulated scales.

This chapter breaks down core principles behind ingredient selection for common concerns. You'll learn why acne needs a multi-pronged attack on oil, bacteria, and inflammation; why aging skin thrives on collagen support and antioxidant defense; and why hyperpigmentation calls for both melanin inhibition and rigorous sun protection. You'll also discover why sensitive skin and eczema require restraint rather than aggression, and how psoriasis benefits from more active interventions.

By understanding how ingredients interact with your skin's biology, you'll be equipped to build smarter, more effective routines, ones that respect your skin's needs while delivering real results.

Skin Barrier Health: Everyone's Foundation

Your skin is in a constant state of renewal. Every day, thousands of cells are born, mature, and shed in a silent cycle known as skin turnover, a process that's essential for maintaining a resilient, functional barrier. This natural rhythm ensures that your skin barrier can defend against environmental stressors, retain moisture, and recover from damage. But when turnover is disrupted, whether by harsh products, inflammation, or age, the skin barrier suffers.

Understanding skin turnover isn't just about knowing how often your skin renews itself. It's about recognizing how this process connects to everything from hydration and sensitivity to acne and aging. A healthy barrier depends on balanced turnover, and balanced turnover depends on the choices you make: the ingredients you apply, the routines you follow, and the stressors you avoid.

Skin cells begin their journey in the basal layer of your epidermis. As discussed in the first chapter, these cells, keratinocytes born in the stratum basale layer, migrate upward to the surface of your skin over 28 to 45 days, taking longer as we age. As they travel, the keratinocytes transform into protective corneocytes, which are the dead, flattened cells that form your skin's outermost protective layer, the stratum corneum. The stratum corneum sheds or flakes off, which ends your skin's natural and continuous renewal process.

How well your skin renewal process works determines how your skin looks and feels. It affects your skin's texture, clarity, and its ability to heal from damage. As it turns over more slowly with age, it shows as dull, dry, and rough skin with increased fine lines and wrinkles.

Over-exfoliation and aggressive treatments can accelerate skin cell turnover beyond what's healthy. This creates a cascade of problems. First comes irritation with redness, stinging, and/or increased sensitivity. Your skin may become dry and flaky as the protective barrier thins. Paradoxically, you might also experience breakouts as oil glands work to overcompensate for your skin's distress.

Aging is a big factor in your skin's exfoliation tolerance. As we age, our bodies become less efficient at creating cellular components, including new skin for skin turnover. The lower layers don't make new cells as quickly, and the protective top layer gets worn down. This is why our skin naturally becomes thinner as we age.

The declining skin turnover rate means dead cells accumulate on the surface. This can lead to a dull, rough, and uneven complexion, and can also contribute to clogged pores, breakouts, and dryness. Exfoliation is important to facilitate skin turnover to keep a buildup of dead cells at bay, however aging skin is more fragile.

If you have a thinning skin barrier due to damage or normal aging, it's especially important to start slowly with exfoliating ingredients, using lower active ingredient concentrations and spacing treatments further apart. A thinning barrier needs more time to recover between exfoliation sessions, and gentle, consistent care will deliver better results than aggressive treatments that damage a-thinning skin barrier.

Whether you're dealing with over-exfoliation damage, age-related barrier thinning, or simply want to maintain optimal skin health, understanding your skin turnover helps you understand skin's ability to heal, hydrate, and defend itself. Healthy skin turnover is key to a healthy skin barrier. Without a strong, resilient skin barrier, even the best ingredients can backfire, causing irritation instead of improvement.

Skin Barrier Stressors

The skin barrier serves as your body's first line of defense, maintaining optimal hydration levels while protecting against environmental irritants and pathogens. When this barrier becomes compromised through over-exfoliation, environmental damage, aging, or underlying skin conditions, it requires targeted repair to restore proper function.

What is a healthy skin barrier? A healthy skin barrier looks smooth, even toned, and naturally luminous. There is no redness, no flaking, and there are no rough patches. It feels supple, soft, and comfortably hydrated, neither tight and dry nor excessively oily. When you apply skincare products, they absorb well without stinging or burning. Your skin generally feels comfortable and resilient rather than reactive or fragile.

What is a compromised skin barrier? When your skin barrier is compromised (damaged), it is less effective at retaining moisture and protecting against harmful external factors. It can look dull, rough, or uneven, sometimes accompanied by visible signs like persistent redness, flaking, dry patches, or breakouts. Additional symptoms include new sensitivity to products that previously caused no issues, skin irritation, areas of excessive dryness, and unexpected oiliness as the skin overcompensates for moisture loss.

Recognizing what damages your skin barrier helps you avoid common stressors while building more effective routines. While this book is about skincare ingredients, understanding the factors that affect your skin health and how they work together helps you define a more tailored approach for maintaining (or recovering) your skin barrier health.

Over-Exfoliation

Excessive use of chemical exfoliants (lactic acid, glycolic acid, etc.) or physical scrubs can thin and even disrupt your skin's protective layer. When over-exfoliation has damaged your skin barrier, it is more vulnerable to environmental damage, and you are more likely to see premature signs of aging. To make matters worse, if the inflammation that accompanies over-exfoliation becomes chronic, it can trigger melanin overproduction, leading to blotchy looking skin with dark spots.

The good news? Over-exfoliation is entirely preventable once you recognize your skin's limits. All you need to do is pay attention to your skin's signals. Persistent tightness, stinging, unusual redness, breakouts, a burning sensation after applying products, skin that suddenly feels rough, or skin that looks shiny without being oily, patchy skin dryness or flaking are all signs of a distressed skin barrier that can be caused by over-exfoliation.

Remember too that your skin's exfoliation tolerance can fluctuate with stress, weather changes, and hormonal shifts. When in doubt, just dial back on exfoliating active ingredients and focus on gentle hydration and barrier repair until your skin feels calm and comfortable again. That might be all you need to do.

Strong Surfactants

Surfactants like sodium lauryl sulfate (SLS) can clean too aggressively. They strip essential lipids along with dirt and debris, leaving skin alkaline and vulnerable. That "squeaky clean" feeling often indicates barrier damage rather than thorough cleansing. Your skin should feel clean but comfortable after washing, never tight or stripped. The right cleanser cleans effectively while respecting barrier integrity.

UV Radiation

Most adults need 5-30 minutes of sunlight exposure without sunscreen on most days of the week to produce the vitamin D their bodies need. However, UV damage can occur with as little as 15 to 30 minutes of unprotected sun exposure, depending on factors like skin type, UV index, time of day, and geographic location.

Using a good mineral barrier sunscreen of SPF 30 or better protects your skin from excess sun exposure, and this should be a whole health priority. UV rays don't just cause direct damage; they also amplify the harmful effects of pollutants. Protecting your skin from excessive UV rays is a top requirement for healthy skin at any age.

Environmental Assaults

Your skin faces constant environmental assault. Air pollution, including particulate matter and ozone, can penetrate your skin. This can cause oxidative stress and accelerate aging, leading to wrinkles and pigmentation. Long-term exposure to polluted air may also aggravate skin conditions like acne and eczema by disrupting the skin barrier function. Antioxidant skincare ingredients help neutralize free radicals and combat oxidative stress.

- **Seasonal Changes:** Most people need to make at least some seasonal skincare adjustments. Temperature and humidity fluctuations can stress your skin's barrier. Your skin's proneness to dryness, irritation, sensitivity, and oil production is likely to fluctuate, especially if the season change includes extreme heat or cold with low or high humidity.

- **General Health:** Making your health a high priority is a big commitment. Sometimes life gives you other priorities, and time and energy just aren't there. It's important to appreciate that your health affects your skin's function - and vice versa. Hormonal fluctuations can alter sebum production and skin thickness. Chronic stress elevates cortisol, impairing barrier function and slowing healing. Nutritional deficiencies compromise your skin's ability to maintain and repair itself. Genetic variations also play a role.

Skin Barrier Support Ingredients

Skin barrier repair ingredients strengthen the skin's natural defenses by restoring hydration, replenishing lipids, and calming inflammation. They are especially beneficial for sensitive, compromised, or reactive skin that needs gentle support rather than aggressive treatment.

Skin barrier repair ingredients focus on strengthening the skin's natural defenses, calming irritation, and restoring healthy hydration. They are especially important for sensitive, reactive, or compromised skin that needs protection and support rather than aggressive treatment. Hydration support works through three primary mechanisms:

Humectants - Draw moisture into the skin in adequately humid environments
Emollients - Smooth and soften skin
Occlusives - Seal in hydration and protect the skin barrier

Multi-function ingredients address multiple skin concerns simultaneously. The following are ingredients you'll most likely find in skincare formulations that can deliver hydration, anti-inflammatory support, and lipid replenishment.

Multi-Function Barrier Repair Ingredients

The following versatile multi-function ingredients address multiple skin barrier repair mechanisms simultaneously. They work synergistically to hydrate, soothe inflammation, restore lipids, and promote healing, all while being gentle enough for sensitive or damaged skin. By targeting several aspects of barrier dysfunction at once, they form the foundation of effective skincare routines, fostering a resilient, healthy skin barrier with minimal irritation.

- **Allantoin**: Soothing and anti-inflammatory agent that promotes skin barrier healing and reduces irritation without any sensitization risk, ideal for reactive skin.
- **Aloe Vera**: Hydrating, anti-inflammatory, and wound-healing properties.
- **Beef Tallow**: Rich in fatty acids and fat-soluble vitamins, it closely mimics natural human lipids to deeply hydrate and soothe irritation.
- **Centella Asiatica (including Madecassoside)**: Rich in compounds that calm irritation, stimulate repair, and improve skin resilience.
- **Ceramides**: Essential skin-identical lipids that strengthen the barrier, retain moisture, and reduce irritation while restoring barrier integrity.
- **Marula Oil**: A deeply moisturizing oil, rich in antioxidants and vitamins that support elasticity and repair.
- **Niacinamide (Vitamin B3)**: Calms inflammation, reduces sebum output, improves skin barrier lipids, and helps fade pigmentation.
- **Panthenol (Provitamin B5)**: Deeply hydrates and helps reduce redness and inflammation while supporting healing processes.
- **Rosehip Oil**: Rich in essential fatty acids and vitamins A and C, supports regeneration and improves texture.
- **Squalane**: Lightweight emollient that mimics natural skin oils, restoring suppleness without clogging pores or heaviness.

- **Urea (≤5%)**: Both hydrates and provides gentle exfoliation to soften rough or flaky skin caused by barrier dysfunction.

Single-Purpose Barrier Repair Ingredients

The skin barrier acts as a critical shield against environmental stressors, and its repair requires targeted ingredients that focus on specific dysfunctions, such as dehydration, inflammation, or lipid depletion. These single-purpose ingredients excel at addressing individual barrier repair mechanisms - whether through hydration, soothing, or structural reinforcement - making them ideal for strategic use in skincare routines. When combined with multi-functional actives, they create a synergistic, tailored approach to restore and maintain a robust, healthy skin barrier.

- **Avocado Oil**: Deeply moisturizing oils rich in antioxidants and vitamins that support elasticity and repair.
- **Bisabolol**: Chamomile extract that reduces redness and soothes sensitive or reactive skin without sensitization risk.
- **Cholesterol**: A skin-identical lipid that directly supports barrier repair by restoring lipid matrix integrity, often paired with ceramides and fatty acids.
- **Colloidal Oatmeal**: Relieves itch, calms inflammation, and reinforces the barrier with proven anti-inflammatory properties.
- **Glycerin**: A foundational humectant that draws moisture into the skin and enhances barrier resilience, with an excellent safety profile.
- Hyaluronic Acid: Hydrates by holding up to 1,000 times its weight in water, suitable even for sensitive compromised skin.
- **Jojoba Oil**: Closely resembles human sebum, balancing oil production while softening and protecting barrier function.
- **Mango Butter**: Rich emollient that provides intensive moisture and barrier protection with natural vitamins A and C for enhanced skin softening and repair.
- **Polyglutamic Acid**: A newer humectant that provides moisture retention for intensive hydration.
- **Shea Butter**: Occlusive emollient packed with nourishing fats and antioxidants that seal in moisture effectively.

An important note on humectants: While humectants like glycerin and hyaluronic acid are excellent for drawing moisture into the skin, they can have the opposite effect in dry, arid climates. Without enough humidity in the air, these ingredients can pull water from the deeper layers of your skin instead, leading to increased dryness. In low humidity environments they need to be paired with occlusives to lock in moisture.

If you have a compromised skin barrier, you need to prioritize skin repair overactive treatment until your skin barrier function is restored. A healthy barrier is more tolerant of and responds better to active ingredients. Successful barrier repair requires patience and consistency. Restoring a distressed skin barrier can take 4 to 8 weeks of focused care.

Acne: A Multi-Factor Challenge

Acne involves four key processes: excess sebum production, abnormal cell shedding that clogs pores, bacterial proliferation, and inflammation. Because these factors interact and amplify each other, effective treatment requires addressing multiple causes simultaneously rather than targeting just one symptom.

Understanding this complexity helps explain why single ingredient approaches often fail while comprehensive routines succeed. The most effective treatments combine ingredients that work through complementary mechanisms, from gentle options like niacinamide for sensitive skin to more potent treatments like retinoids for stubborn breakouts.

The following are some of the most common ingredients in skincare products that are especially effective for acne-prone skin. They work by exfoliating and clearing pores, reducing inflammation, regulating oil production, and reinforcing the skin's barrier so it can better defend itself.

Multi-Function Acne Ingredients

The following multi-functional skincare ingredients tackle multiple acne triggers simultaneously, providing comprehensive benefits like exfoliation, sebum regulation, and inflammation reduction. These gentler alternatives to single-purpose actives form a robust foundation for acne-prone skincare routines, delivering balanced results with minimal irritation.

- **Adapalene**: A specific retinoid that is available over-the-counter at 0.1% with anti-inflammatory, comedolytic, and barrier-supporting properties.
- **Azelaic Acid**: Gently exfoliates and brightens while reducing acne bacteria, calming redness, and soothing inflammation.
- **Lactic Acid**: A gentle AHA that exfoliates, hydrates, and has mild antibacterial properties, suitable for sensitive acne-prone skin.
- **Mandelic Acid**: Gently exfoliates, fights bacteria, and suits sensitive acne-prone skin.
- **Niacinamide**: Calms inflammation, reduces sebum output, improves skin barrier lipids, and helps fade pigmentation.
- **Retinoids**: Normalize cell turnover, prevent clogged pores, and reduce inflammation. Options range from gentle retinol to prescription-strength tretinoin.
- **Salicylic Acid (BHA)**: Penetrates pores to clear buildup, reduces inflammation, and dissolves dead skin cells at 0.5-2% concentrations.
- **Tea Tree Oil**: Natural antimicrobial effective against acne bacteria with anti-inflammatory, and sebum-regulating effects, best used as a spot treatment or in diluted formulations.
- **Zinc**: Anti-inflammatory mineral that reduces oil production and supports healing, available in both topical and supplement forms.

Single-Purpose Acne Ingredients

These single-purpose skincare ingredients focus on specific acne mechanisms, such as exfoliation, oil control, or bacterial reduction, delivering precise results with minimal irritation. They're commonly incorporated into cleansers, toners, and spot treatments and often excel when paired with multi-functional actives.

- **Alpha Hydroxy Acids:** Glycolic acid, lactic acid, mandelic acid, citric acid, malic acid, and tartaric acid are AHAs widely used in acne treatments, especially non-inflammatory acne, that exfoliate to prevent clogged pores.
- **Bakuchiol:** A gentler pregnancy-safe retinol alternative that accelerates cell turnover and may reduce melanin production, providing anti-aging benefits alongside hyperpigmentation treatment.
- **Benzoyl Peroxide:** Antimicrobial that kills bacteria and clears existing breakouts, effective at 2.5-10% (higher isn't necessarily better).
- **Centella Asiatica:** Soothes irritated skin and supports healing.
- **Clay Minerals (Kaolin, Bentonite):** Absorb excess oil and draw impurities from pores without over-stripping.
- **Sulfur:** Natural mineral that reduces oiliness and fights bacteria with minimal side effects.
- **Witch Hazel:** An astringent that reduces oil and soothes inflammation, often used in toners for acne-prone skin.

Acne Power Combo: Niacinamide and Zinc

Niacinamide and zinc are frequently paired in acne formulations because their complementary actions create a balanced, targeted approach to managing acne and oily skin. Niacinamide, a versatile derivative of vitamin B3, addresses multiple concerns simultaneously: reducing inflammation, regulating sebum production, strengthening the skin barrier, and fading post-acne marks.

Zinc, often included as zinc oxide or zinc PCA, provides focused anti-inflammatory and oil-controlling benefits, helping to calm active breakouts and reduce shine. Together, they form a synergistic duo: niacinamide broadens the therapeutic scope, while zinc delivers precision in oil regulation and inflammation control. This pairing provides multi-layered benefits without overwhelming skin, making it ideal for sensitive or acne-prone users.

Beyond Topical Care

Acne management extends beyond the products you apply to your skin. A comprehensive approach addresses lifestyle factors and environmental triggers that can significantly impact breakout frequency and severity. Acne-prone skin is especially vulnerable to bacterial buildup, which can trigger inflammation and cause new breakouts. Taking steps to protect affected skin from bacterial contamination is essential.

- **Dietary Factors:** Common dietary culprits include high-glycemic foods like refined sugars and processed carbohydrates. These foods can spike blood sugar and insulin levels, affect hormones, and/or increase sebum production. Some people find that gluten, soy, dairy products, or certain proteins trigger inflammatory responses that worsen acne. Identifying specific triggers often requires careful observation or a restrictive elimination diet.

- **Clean Bedding Protocol:** Change pillowcases every 1-2 nights, as they collect oil, dead skin cells, hair products, and bacteria throughout the night. Consider using antimicrobial pillowcase materials or having multiple pillowcases to rotate. Use hot water (at least 140°F) and fragrance-free detergent to effectively eliminate bacteria from bedding.

- **Fresh Towels and Washcloths:** Use a clean washcloth for each cleansing session and a fresh towel for drying. Damp towels become breeding grounds for bacteria within hours. Dedicate specific towels for use on your face and keep them separate from body towels to prevent cross-contamination from body oils and products.

- **Hand and Hair Hygiene:** Avoid touching your face throughout the day, as hands transfer bacteria, oil, and environmental pollutants that can clog pores. Keep hair clean and off the face, especially while sleeping. Hair products containing oils, silicones, or heavy conditioning agents can migrate to pillowcases and subsequently to facial skin, triggering breakouts along the hairline and forehead.

- **Phone and Surface Cleanliness:** Every so often there's a news story about cell phones harboring more germs than a toilet seat. Regularly disinfect phones, glasses, and other items that contact your face. These surfaces accumulate bacteria and oil throughout the day and can reintroduce contaminants to freshly cleansed skin.

- **Light Therapy Options:** LED blue light therapy (405-420 nm) effectively treats mild to moderate acne by targeting bacteria and reducing inflammation. While generally safe, it requires consistent sessions two to three times weekly and works best as part of a comprehensive treatment plan that includes proper hygiene protocols.

Building Your Acne Strategy

Successful acne treatment requires patience and consistency. Most routines take 6 to 12 weeks to show significant improvement, and the key is maintaining gentle, consistent care while avoiding harsh scrubbing or picking.

Remember that acne treatments can cause dryness and irritation, making barrier support paramount. A compromised barrier can make acne worse, so gentle care remains essential even when targeting breakouts aggressively. Use non-comedogenic moisturizers with ingredients like ceramides to maintain skin health during treatment.

If over-the-counter treatments haven't improved your skin after 12 weeks of consistent use, consider consulting licensed skincare specialist who specializes in acne treatment. They can perform extractions, deliver acne-specific treatment protocols, guide product selection, and can provide professional products with ingredients the FDA says are too potent for over-the-counter sales.

If you have severe or persistent acne that is significantly impacting your quality of life, consider consulting a dermatologist who can prescribe stronger retinoids, prescribe oral medications, and/or recommend in-office procedures.

Rosacea: Managing Chronic Inflammation

Rosacea is a persistent inflammatory skin condition characterized by redness, visible blood vessels, and sometimes small bumps or pustules on the face. It typically affects the cheeks, nose, forehead, and chin, leaving skin feeling hot, irritated, and sensitive. While rosacea can resemble acne, it is fundamentally different; it stems from chronic inflammation and requires ongoing management, not spot treatment.

This complex condition is influenced by a combination of genetic predisposition, environmental triggers, immune system dysregulation, and vascular instability. Although its exact cause remains unclear, research suggests that an overactive immune response and impaired skin barrier function play central roles. People with fair skin are more susceptible, though rosacea can affect individuals of all skin tones.

Effective rosacea management requires a two-pronged approach: identifying and avoiding personal triggers, and using gentle, anti-inflammatory skincare that supports the skin barrier. This is not about aggressive correction - it's about creating a stable environment in which chronically inflamed skin can heal and remain calm.

Rosacea triggers are highly individual, but common culprits include UV exposure, extreme temperatures, spicy foods, alcohol, hot beverages, certain medications, and emotional stress. Even everyday habits, little things like washing your face with hot water, can provoke flare-ups. Keeping a trigger diary can help identify your personal patterns.

Skincare Ingredients to Avoid

Rosacea-prone skin is highly reactive and requires careful ingredient selection. Some ingredients that are beneficial for many skin types will inflame rosacea-prone skin. It is important to avoid:

- Retinoids and retinols
- AHAs and BHAs
- Vitamin C (unless formulated for sensitive skin)
- Alcohol-based products
- Harsh exfoliants
- Fragrances and essential oils

Multi-Function Rosacea Ingredients

The following multi-functional ingredients address multiple rosacea triggers simultaneously, including inflammation, redness, and barrier dysfunction, while remaining soothing for sensitive skin. These versatile actives form the cornerstone of comprehensive rosacea routines, delivering balanced benefits to calm flare-ups and support long-term skin health with minimal irritation.

- **Azelaic Acid:** Gently exfoliates and brightens while reducing acne bacteria, calming redness, reducing inflammation.
- **Beta-Glucan:** Soothes inflammation, hydrates, and supports barrier repair, commonly used in sensitive skin formulations.
- **Centella Asiatica:** Calms inflammation, promotes healing, and provides comprehensive barrier support with proven anti-inflammatory compounds.
- **Licorice Root Extract:** Inhibits tyrosinase production, reduces redness, and soothes underlying irritation while providing sustained anti-inflammatory benefits.
- **Niacinamide:** Calms inflammation, reduces sebum output, improves skin barrier lipids, and helps fade pigmentation.
- **Sulfur:** Addresses inflammation and bacterial growth in papulopustular rosacea, providing multiple benefits with gentle action.
- **Tranexamic Acid:** Reduces redness, calms inflammation, interrupts melanin production pathways, and addresses stubborn discoloration.

Single-Purpose Rosacea Ingredients

Managing rosacea requires targeted interventions to address specific symptoms, such as inflammation, sensitivity, or barrier impairment, without causing irritation. The following single-purpose ingredients focus on individual rosacea mechanisms, providing gentle, precise relief for redness, reactivity, or hydration deficits. Commonly used in skincare, these ingredients shine in strategic combinations with multi-functional actives, creating tailored, soothing routines to support sensitive, rosacea-prone skin.

- **Aloe Vera:** Hydrates and soothes, widely used in rosacea care for its gentle, cooling properties.
- **Bisabolol:** Chamomile extract that reduces redness and soothes sensitive or reactive skin without sensitization risk.
- **Ceramides:** Restore compromised barrier function that often underlies rosacea sensitivity.
- **Chamomile Extract:** Time-tested anti-inflammatory that soothes irritated skin without sensitization risk.
- **Colloidal Oatmeal:** Provides immediate anti-inflammatory relief and gentle barrier support.
- **Feverfew Extract:** Reduces both sensitivity and inflammatory responses, particularly effective for reactive skin.
- **Green Tea Extract:** Rich in polyphenols that provide anti-inflammatory benefits and antioxidant protection.
- **Hyaluronic Acid:** Provides essential hydration without any irritation potential, suitable for the most sensitive skin; requires adequate humidity.
- **Metronidazole:** A prescription topical antibiotic with proven anti-inflammatory properties for moderate to severe rosacea.
- **Oat Extract:** Similar to colloidal oatmeal but sometimes used in lighter formulations for soothing.
- **Squalane:** Lightweight emollient that mimics natural skin oils, restoring suppleness without clogging pores or heaviness.

- **Zinc Oxide:** Provides physical sun protection with mild anti-inflammatory and antimicrobial properties.

Beyond Topical Care

Rosacea is a chronic inflammatory condition deeply influenced by lifestyle factors, environmental triggers, and systemic health. A comprehensive approach focuses on trigger identification, inflammation reduction, and gentle protection strategies.

- **Gut-Skin Connection:** Imbalances can contribute to systemic inflammation that worsens rosacea symptoms. Consider incorporating probiotics, prebiotics, and anti-inflammatory foods as complementary strategies for managing this condition from within.

- **Trigger Identification and Avoidance:** Rosacea triggers are highly individual, so identifying personal triggers is critical for effective management. Common triggers include spicy foods, hot beverages, alcohol (especially red wine), extreme temperatures, stress, certain skincare ingredients, and UV exposure. Keeping a detailed trigger diary noting foods consumed, activities, weather conditions, stress levels, and products used - alongside flare-up timing - can help identify specific patterns.

- **Gentle Sun Protection:** UV exposure is one of the most common rosacea triggers, making daily broad-spectrum SPF 30+ essential. Choose mineral sunscreens with zinc oxide or titanium dioxide, as chemical sunscreens can irritate sensitive rosacea skin. Seek shade during peak hours, wear wide-brimmed hats, and remember that UV rays reflect off snow, water, and concrete, potentially triggering flares even in shaded areas.

- **Temperature Management:** Extreme temperatures - both hot and cold -can trigger vasodilation and worsen rosacea symptoms. Avoid saunas, hot tubs, and very hot showers. Also avoid exercise and activities that cause overheating. In cold weather, protect your face with a scarf or mask. When exercising, choose cooler environments and use fans to prevent overheating that can trigger flushing.

When to See a Professional

A licensed skincare specialist can help by providing treatments and products with concentrated, targeted ingredients unavailable in retail stores. A dermatologist can prescribe treatments such as topical antibiotics, high dose azelaic acid, or oral medications that provide additional support for moderate to severe cases. Early intervention often prevents progression to more advanced stages of rosacea.

Building Your Rosacea Strategy

Successful rosacea management depends on consistency, restraint, and a focus on inflammation control. Unlike acne, rosacea doesn't respond well to aggressive treatment. Instead, calming the skin and preventing triggers is the long-term goal.

With rosacea, less is often more. A minimal routine with proven anti-inflammatory ingredients typically works better than complex regimens that risk overwhelming sensitive skin.

Hyperpigmentation: Interrupting Melanin Production

Melanin is an important antioxidant and free radical scavenger. It's produced by melanocytes in the stratum basale layer of your skin and is naturally reddish-yellow (pheomelanin) or brownish-black (eumelanin). Hyperpigmentation occurs when melanocytes overproduce melanin because of excessive UV exposure, inflammation from acne, or hormonal changes.

Targeted Skincare Ingredients

Successful hyperpigmentation management requires a three-pronged approach: prevention of UV aggravation with excellent sun protection, inhibition of new melanin production, and acceleration of cell turnover to fade existing darkness. This isn't about bleaching your skin - it's about normalizing melanin production and helping your skin shed excess pigment naturally.

Daily broad-spectrum sunscreen with at least SPF 30 is non-negotiable. UV exposure not only darkens existing spots but also triggers new pigmentation, undermining any progress from treatment products. Products with glycolic or salicylic acid often provide gentle exfoliation that helps pigmented surface cells shed to enhance the penetration of brightening ingredients.

Multi-Function Hyperpigmentation Ingredients

Hyperpigmentation, driven by overactive melanin production, requires a multifaceted approach to fade existing discoloration and prevent new spots. These multi-functional ingredients tackle multiple pathways - such as melanin synthesis, cell turnover, and inflammation - while providing additional skin benefits like barrier support and antioxidant protection. Gentle yet effective, they form the cornerstone of comprehensive brightening routines, addressing post-inflammatory hyperpigmentation, melasma, and sun-induced spots with minimal irritation for lasting clarity.

- **Azelaic Acid**: Gently exfoliates and brightens while reducing acne bacteria, calming redness, and reducing inflammation.
- **Bakuchiol**: A gentler pregnancy-safe retinol alternative that accelerates cell turnover and may reduce melanin production, providing anti-aging benefits alongside hyperpigmentation treatment.
- **Licorice Root Extract**: Inhibits tyrosinase production, reduces redness, and soothes underlying irritation while providing sustained anti-inflammatory benefits.
- **Mandelic Acid**: Larger molecule AHA gentler than glycolic acid that provides exfoliation and antibacterial properties for sensitive skin.
- **Niacinamide**: Blocks melanin transfer between cells, supports barrier function, and reduces inflammation, with optimal benefits at 4-5% concentration.

- **Retinoids:** Accelerate cell turnover to fade existing spots while preventing new melanin formation through normalized cell renewal processes.
- **Tranexamic Acid:** Reduces redness, calms inflammation, interrupts melanin production pathways, and addresses stubborn discoloration.
- **Vitamin C:** Inhibits melanin production, provides antioxidant protection, and brightens skin, with L-ascorbic acid being most potent at 10-20% concentration.

Important Safety Note: While hydroquinone was once the gold standard for pigmentation treatment, safety concerns - including potential carcinogenic properties and the risk of exogenous ochronosis (a paradoxical blue-black discoloration) have led to restrictions in many countries. Fortunately, newer ingredients provide effective alternatives with better safety profiles.

Single-Purpose Hyperpigmentation Ingredients

Addressing hyperpigmentation requires targeted strategies to inhibit melanin production or remove pigmented cells without exacerbating sensitivity. These single-purpose ingredients focus on specific mechanisms, such as tyrosinase inhibition or exfoliation, delivering precise results for fading dark spots. Widely used in serums, masks, and spot treatments, they excel when combined with multi-functional actives, creating synergistic routines that promote even-toned skin while minimizing irritation.

- **Alpha-Arbutin:** Natural hydroquinone derivative that prevents melanin formation without hydroquinone's irritation potential, effective at 1-2% concentration.
- **Ascorbyl Glucoside:** Stable vitamin C derivative that converts to active form in skin, providing brightening benefits with reduced irritation potential.
- **Cysteamine:** Emerging ingredient that works through multiple pathways to reduce pigmentation, showing promise even for stubborn melasma.
- **Glycolic Acid:** Removes pigmented surface cells through exfoliation and accelerates cell turnover to fade existing discoloration.
- **Kojic Acid:** Natural tyrosinase inhibitor derived from fungi, effectively blocks melanin production with proven brightening results.
- **Ferulic Acid:** Stabilizes and enhances vitamin C effectiveness, extending its brightening benefits.
- **Mulberry Extract:** Natural tyrosinase inhibitor that prevents melanin formation with minimal irritation risk.
- **N-Acetyl Glucosamine:** Inhibits melanin production and enhances exfoliation, often paired with niacinamide for synergy.
- **Phytic Acid:** Gently exfoliates and inhibits melanin, common in natural formulations.
- **Resorcinol Derivatives (e.g., Hexylresorcinol):** Inhibit tyrosinase and are gentler alternatives to hydroquinone, gaining popularity in brightening products.
- **Rice Extract:** Contains phytic acid and other compounds that gently brightens skin while providing antioxidant protection.
- **Salicylic Acid:** Exfoliates to remove pigmented cells and is effective for PIH, especially in acne-prone skin, commonly used at 0.5-2%.

- **Vitamin E:** Works synergistically with vitamin C to enhance brightening effects while providing antioxidant protection against new damage.

Building Your Hyperpigmentation Strategy

Fading hyperpigmentation demands patience, consistency, and a strategic approach, as visible results typically emerge after 8 to 12 weeks of diligent care. The cornerstone of success lies in combining gentle, targeted ingredients with non-negotiable daily broad-spectrum sunscreen (SPF 30 or higher) to prevent UV-induced pigment worsening.

Begin with one or two well-tolerated actives, such as niacinamide or vitamin C, to inhibit melanin production and promote skin clarity, gradually introducing additional ingredients as your skin adjusts. Overly aggressive treatments can trigger inflammation and worsen pigmentation, so prioritize gentle, consistent routines to safely and effectively achieve an even-toned complexion.

Melasma: The Stubborn Pigmentation Condition

Melasma stands apart from other forms of hyperpigmentation due to its complex hormonal triggers and notorious resistance to treatment. This condition appears as symmetrical patches of discoloration, typically on the forehead, cheeks, bridge of the nose, upper lip, and chin. Unlike post-inflammatory hyperpigmentation from acne or sun damage, melasma involves intricate interactions between melanocytes, hormones, UV exposure, and vascular factors that make it particularly challenging to address.

Hormonal fluctuations from pregnancy, oral contraceptives, and hormone replacement therapy increase melanocyte sensitivity to UV radiation. This heightened sensitivity means that even minimal sun exposure can trigger significant pigment production in those predisposed, often undoing months of treatment progress.

Melasma exists in three distinct forms based on pigment depth, each requiring different treatment approaches:

- **Epidermal** melasma appears brown and well-defined, responding better to topical treatments.
- **Dermal** melasma appears as grayish-brown pigmentation with poorly defined borders and proves more resistant to surface treatments.
- **Mixed** melasma combines both types, requiring comprehensive strategies that address multiple pigment depths.

Recent research has identified a significant link between hyperinsulinemia (elevated blood insulin) and melasma development. High insulin levels stimulate melanocyte activity through cross-activation of insulin-like growth factor (IGF-1) receptors. This metabolic connection helps explain why melasma often emerges during hormonal changes and why topical treatments alone frequently fall short.

Addressing insulin resistance through dietary modifications, particularly reducing refined carbohydrates and emphasizing antioxidant-rich foods, can significantly improve melasma outcomes when combined with topical therapies. This internal approach targets one of melasma's root causes rather than just surface symptoms.

Targeted Skincare Ingredients

Daily broad-spectrum sunscreen with at least SPF 30 is non-negotiable for any hyperpigmentation, especially melasma, because UV exposure triggers new pigmentation. Gentle exfoliation with acids like glycolic or salicylic acid helps remove pigmented surface cells and enhances the penetration of brightening ingredients.

Multi-Function Melasma Ingredients

Melasma, a complex condition driven by hormonal, UV, and inflammatory triggers, demands versatile ingredients that address multiple pathways while remaining gentle for long-term use. These multi-functional actives target melanin production, inflammation, and skin barrier health, providing comprehensive benefits for fading stubborn discoloration and preventing flare-ups. They form the backbone of effective melasma management, delivering safe, sustained results with minimal irritation.

- **Azelaic Acid:** Gently exfoliates and brightens while reducing acne bacteria, calming redness, and reducing inflammation.
- **Bakuchiol:** A gentler pregnancy-safe retinol alternative that accelerates cell turnover and may reduce melanin production, providing anti-aging benefits alongside hyperpigmentation treatment.
- **Licorice Root Extract:** Inhibits tyrosinase production, reduces redness, and soothes underlying irritation while providing sustained anti-inflammatory benefits.
- **Niacinamide:** Calms inflammation, reduces sebum output, improves skin barrier lipids, and helps fade pigmentation.
- **Phytic Acid:** Gently exfoliates and inhibits melanin, common in natural formulations.
- **Retinoids:** Accelerate cell turnover to fade existing spots while preventing new melanin formation through normalized cell renewal processes.
- **Tranexamic Acid:** Reduces redness, calms inflammation, interrupts melanin production pathways, and addresses stubborn discoloration.
- **Tretinoin:** A prescription-strength retinoid derived from vitamin A that accelerates cell turnover and enhances penetration of other actives, often essential for stubborn melasma but requires careful monitoring.
- **Vitamin C:** Safe during pregnancy, effective for melasma prevention, and provides essential antioxidant protection against UV-triggered flare-ups.

Important Safety Note: While hydroquinone was once the gold standard for pigmentation treatment, including melasma, safety concerns - including potential carcinogenic properties and the risk of exogenous ochronosis (paradoxical blue-black discoloration) have led to restrictions in many countries. Fortunately, newer ingredients provide effective alternatives with better safety profiles.

Single-Purpose Melasma Ingredients

Melasma's persistent nature requires targeted ingredients that focus on specific mechanisms, such as melanin inhibition or cell turnover, to fade discoloration effectively. These single-purpose targeted actives deliver precise results. Because of melasma's complexity and potential side effects, using them effectively often requires some professional guidance. Ideal for combination with multi-functional ingredients, they enhance comprehensive melasma routines, particularly when tailored for pregnancy-safe or sensitive skin applications, ensuring safe and strategic management of stubborn pigmentation.

- **Alpha-Arbutin**: Gentler hydroquinone alternative that works effectively in combination therapy without harsh side effects.
- **Cysteamine**: Newer agent showing significant promise for melasma treatment through multiple anti-pigmentation pathways.
- **Glycolic Acid**: Enhances penetration of other actives while providing gentle exfoliation to remove pigmented cells.
- **Hexylresorcinol**: Synthetic compound with proven effectiveness specifically for melasma treatment.
- **Kojic Acid**: Particularly effective for melasma when combined with other agents, providing targeted tyrosinase inhibition.
- **N-Acetyl Glucosamine**: Inhibits melanin production and enhances exfoliation, often paired with niacinamide for synergy.
- **Rumex Occidentalis Extract**: Plant extract specifically studied and developed for melasma management.
- **Salicylic Acid**: Exfoliates to remove pigmented cells, effective for melasma-related PIH, and pregnancy-safe at low concentrations (0.5-2%).

Building Your Melasma Strategy

Managing melasma requires unwavering commitment, patience, and a tailored approach, as visible improvements often take 3 to 6 months of consistent care, with ongoing maintenance to prevent recurrence. Driven by hormonal fluctuations, UV exposure, and inflammation, melasma can resurface during pregnancy, contraceptive use, or other hormonal shifts, even after successful treatment.

Daily broad-spectrum sunscreen (SPF 30 or higher) is non-negotiable to block UV-triggered flare-ups, while combination therapies, often under dermatological supervision, maximize efficacy. By prioritizing gentle, pregnancy-safe ingredients and consistent routines, significant fading and long-term control of melasma are achievable.

Aging & Photodamage: Supporting Structural Integrity

As we age, our skin naturally loses structural integrity through a gradual decline in collagen, elastin, and natural hydration systems. While this process begins in our twenties with approximately 1% collagen loss per year, the most significant accelerator of visible aging is photodamage.

Photodamage is the cumulative damage from repeated UV exposure that triggers oxidative stress and chronic inflammation. UV radiation doesn't just cause immediate sunburn; it systematically breaks down the proteins that give your skin its strength and elasticity. This leads to wrinkles, sagging, uneven pigmentation, and rough texture that can make skin appear significantly older than chronological age would suggest.

Understanding the Aging Process

By our 30s, the effects of both natural aging and accumulated photodamage become increasingly visible. The skin's natural repair mechanisms begin to slow, making it less efficient at defending skin from daily environmental stressors and UV exposure. Cell turnover rates decrease from every 14 days to a slower cycle of 28 days or more. This deceleration means that damaged cells linger longer on the skin's surface, contributing to a dull, uneven complexion.

In our 40s, the fine lines multiply, and deeper wrinkles can begin to form, especially around the eyes and mouth. Thinning skin becomes noticeably less firm and elastic. Age spots and sun damage become more prominent. Hormonal and structural differences give male and female skin distinctly different aging patterns. Higher testosterone means male skin is approximately 25% thicker, and the male testosterone drop is slower than the dramatic drop in estrogen females experience during menopause. This means males lose about 1% of collagen per year consistently, while female collagen loss accelerates significantly during the first five years after menopause.

The 50s brings increased dryness, thinning, and fragility. Jowls and sagging become apparent as facial volume decreases, while the skin's barrier function weakens. During this time, differing male and female different wrinkle patterns become more pronounced. Males tend to develop deeper forehead wrinkles (brow furrows) and more prominent under-eye wrinkles while females typically see wrinkles appear first around the mouth (smoker's lines, marionette lines) and lower cheeks.

By our 60s and beyond, significant loss of facial fat and bone density creates pronounced volume loss and deeper wrinkles in both sexes. Skin becomes markedly thinner and more fragile, with increased bruising and slower wound healing.

The key to healthy, glowing skin as you age is supporting your skin's structural foundation while protecting against damage that accumulates. This means stimulating collagen production, accelerating cell renewal, providing antioxidant protection, and maintaining optimal hydration, all while using gentle methods that don't stress aging skin.

Multi-Function Aging & Photodamage Power Ingredients

Aging and photodamage manifest as wrinkles, loss of elasticity, uneven tone, and compromised barrier function, requiring versatile ingredients that tackle multiple concerns simultaneously. These multi-functional actives stimulate collagen production, enhance cell turnover, provide antioxidant protection, and support skin barrier health, delivering comprehensive anti-aging benefits with balanced results. Forming the foundation of effective skincare routines, they address fine lines, sunspots, and texture changes while minimizing irritation, making them ideal for long-term use in combating visible signs of aging.

- **Azelaic Acid:** Gently exfoliates and brightens while reducing acne bacteria, calming redness, reducing inflammation.
- **Bakuchiol:** A gentler pregnancy-safe retinol alternative that accelerates cell turnover and may reduce melanin production, providing anti-aging benefits alongside hyperpigmentation treatment.
- **Coenzyme Q10 (Ubiquinone):** Supports collagen production, provides antioxidant protection, and reduces oxidative stress, effective for photoaging.
- **Niacinamide:** Calms inflammation, reduces sebum output, improves skin barrier lipids, and helps fade pigmentation.
- **Peptides:** A growing class of ingredients that signal collagen and elastin production while supporting cellular repair, with copper peptides and palmitoyl pentapeptide showing best effectiveness.
- **Retinoids:** Stimulate collagen production, normalize cell turnover, and improve skin texture through proven cellular mechanisms, with options from gentle retinol to prescription tretinoin.
- **Tranexamic Acid:** Reduces redness, calms inflammation, interrupts melanin production pathways, and addresses stubborn discoloration.
- **Vitamin C:** Essential for collagen synthesis while providing antioxidant protection and brightening effects, with L-ascorbic acid being most potent at 10-20% concentration.

Single-Function Aging & Photodamage Support Ingredients Power Ingredients

Addressing specific aspects of aging and photodamage such as cellular damage, moisture loss, or UV-induced degradation requires targeted ingredients that deliver precise results. These single-function actives focus on individual mechanisms, like exfoliation, hydration, or antioxidant defense, to combat fine lines, dullness, and environmental damage. Widely used in serums, moisturizers, and treatments, they excel when paired with multi-functional ingredients, creating synergistic routines that restore youthful skin vitality and protect against further photodamage.

- **Alpha Lipoic Acid:** A potent antioxidant that reduces oxidative stress and improves skin texture, effective for photoaging.
- **Ceramides:** Restore compromised barrier function essential for healthy aging and moisture retention.
- **Exosomes:** Extracellular vesicles that deliver regenerative signals directly to aging cells.
- **Green Tea Polyphenols:** Anti-inflammatory compounds that protect against UV and pollution damage.
- **Growth Factors:** Promote cellular renewal and repair through targeted signaling pathways.
- **Glycolic Acid:** Removes damaged surface cells while stimulating deeper cellular renewal at 5 to 10% concentrations.
- **Hyaluronic Acid:** Plumps and hydrates aging skin while improving overall texture and minimizing fine lines.
- **Idebenone:** Powerful cellular antioxidant that outperforms many traditional options for protection and repair.
- **Lactic Acid:** Gentler AHA option that provides exfoliation plus hydration benefits for mature skin.

- **Squalane:** Lightweight emollient that mimics natural skin oils, restoring suppleness without clogging pores or heaviness.
- **Resveratrol:** Potent antioxidant that specifically protects against photoaging and environmental damage.
- **Vitamin E + Ferulic Acid:** Synergistic combination that enhances and stabilizes other antioxidants like vitamin C.

Photodamage Power Combo Vitamin E + Ferulic Acid

Vitamin E and ferulic acid form a synergistic combination for aging skin by enhancing antioxidant protection and stabilizing each other's effects. Vitamin E, a potent antioxidant, neutralizes free radicals caused by UV and environmental damage, reducing oxidative stress that contributes to wrinkles and photodamage. Ferulic acid, another antioxidant, boosts the stability and efficacy of vitamin E (and often vitamin C in formulations), extending their protective benefits.

Together, they shield skin from collagen degradation, improve barrier function, and reduce inflammation, making them particularly effective in combating fine lines, sunspots, and texture changes associated with aging. This combination is commonly found in serums, providing a targeted defense against photoaging while supporting overall skin health.

Beyond Topical Treatment

Optimizing aging skin extends far beyond the products you apply to your skin. A comprehensive approach that incorporates lifestyle and environmental factors can significantly affect how your skin looks and functions as you age.

- **Sleep and Recovery:** Your body undergoes its most intensive repair processes during quality sleep. Be sure you are getting enough.
- **Clean Eating:** A nutrient-dense, whole-foods diet with adequate protein (we need more as we age) provides the building blocks your skin needs for repair and protection.
- **Exercise:** Weight-bearing exercise with some cardio is critical for skin health as we age. Muscle mass provides skin structural support, and cardio delivers nutrients to skin cells while helping flush away cellular waste products.
- **Stress Management:** Chronic stress elevates cortisol levels, which breaks down collagen and impairs skin barrier function.
- **Environmental Protection:** Use antioxidant serums in the morning, consider air purifiers (including plants) in your living spaces, and gently cleanse skin each evening to remove environmental pollutants that accumulate throughout the day.
- **Professional Treatments:** Professional Treatments: Today's professional treatments and post-procedure topicals address sun damage, stimulate collagen production, support skin regeneration, and improve overall skin function.

Building Your Anti-Aging Strategy

Successful anti-aging treatment requires patience, consistency, and a long-term perspective. Most skincare routines take 8 to 12 weeks to show noticeable improvement in texture and tone, while more significant changes in fine lines and firmness can take 3 to 6 months of consistent use. The key is maintaining gentle, progressive care while protecting your skin barrier.

Remember that anti-aging active ingredients can be sensitizing and drying, making barrier support a priority. Use nourishing moisturizers containing ingredients like ceramides, hyaluronic acid, and peptides to maintain skin health during treatment. A compromised barrier can accelerate aging by increasing inflammation and reducing the skin's ability to retain moisture and repair itself.

Start slowly with active ingredients, particularly retinoids and acids. Begin with lower concentrations 2-3 times per week and gradually increase frequency as your skin builds tolerance. This approach prevents irritation that can set back your progress and cause unnecessary sensitivity.

If over-the-counter treatments haven't provided the improvement you're seeking after 3 to 4 months of consistent use, consider consulting a professional who specializes in anti-aging skincare treatments. They can provide professional skin rejuvenation and corrective treatments as well as more potent professional products that deliver stronger results than consumer products.

Remember that consistency beats intensity; a gentle routine you can maintain daily will always outperform aggressive treatments that you can only tolerate occasionally. The goal is sustainable, long-term skin health that improves or at least does not deteriorate as quickly with time.

Eczema: A Skin Barrier Dysfunction Condition

Eczema (atopic dermatitis) represents one of the most complex skin conditions, involving three interconnected processes: severely compromised barrier function, immune system dysregulation, and chronic inflammation. Unlike other skin concerns that might have single triggers, eczema creates a cycle where each factor amplifies the others, making effective management require a comprehensive approach.

The hallmark of eczema is a defective skin barrier that allows irritants and allergens to penetrate while permitting excessive water loss. This compromised barrier triggers immune responses and inflammation that further damages the barrier, creating a self-perpetuating cycle that requires careful intervention to break.

Environmental and lifestyle factors significantly impact eczema flare-ups, making trigger identification foundational for long-term management. Common culprits include harsh soaps and detergents, synthetic fragrances, extreme temperatures, stress, and certain fabrics like wool or synthetic materials. Food allergies can also contribute, with eggs, dairy, nuts, and wheat being frequent offenders.

The skin microbiome plays a particularly important role in eczema. Affected skin often shows reduced microbial diversity and increased colonization by *Staphylococcus aureus*, which can worsen inflammation and barrier dysfunction. Supporting a healthy skin microbiome becomes an essential component of comprehensive eczema care.

Targeted Skincare Ingredients

Eczema management typically requires professional support. Prescription treatments such as topical corticosteroids and calcineurin inhibitors are important tools for managing moderate to severe flares. Wet wrap therapy and phototherapy can provide significant relief for persistent cases when used under medical supervision.

The goal of skincare ingredients is to support and maintain the improvements achieved through professional treatment while preventing future flare-ups through consistent barrier support. Ceramides are essential for eczema management, as they restore the compromised lipid barrier that characterizes this condition. They prevent water loss while supporting overall barrier integrity, making them the foundation of any effective eczema routine.

Colloidal oatmeal deserves special mention for its gentle versatility; it provides anti-inflammatory and soothing properties while providing mild cleansing benefits, making it ideal for sensitive, eczema-prone skin that cannot tolerate traditional cleansers.

Multi-Function Eczema Power Ingredients

Eczema, characterized by a compromised skin barrier, inflammation, and intense irritation, requires versatile ingredients that address multiple symptoms while remaining gentle for sensitive skin. These multi-functional actives soothe inflammation, restore lipid integrity, hydrate, and promote healing, providing comprehensive support for eczema-prone skin. Commonly found in high-quality, fragrance-free formulations, they form the cornerstone of effective eczema management, delivering balanced relief and long-term barrier repair with minimal risk of sensitization.

- **Allantoin**: Soothing and anti-inflammatory agent that promotes skin barrier healing and reduces irritation without any sensitization risk, ideal for reactive skin.
- **Beta-Glucan**: Soothes inflammation, hydrates, and supports barrier repair, commonly used in sensitive skin formulations.
- **Centella Asiatica**: Soothes inflammation promotes healing, and supports barrier repair, widely used in eczema formulations, especially in K-beauty.
- **Colloidal Oatmeal**: Provides gentle anti-inflammatory and soothing properties while providing mild cleansing benefits, ideal for sensitive eczema-prone skin.
- **Licorice Root Extract**: Inhibits tyrosinase production, reduces redness, and soothes underlying irritation while providing sustained anti-inflammatory benefits.
- **Niacinamide**: Calms inflammation, reduces sebum output, improves skin barrier lipids, and helps fade pigmentation.
- **Panthenol**: Hydrates, reduces inflammation, and promotes healing, a multi-functional staple for eczema-prone skin.

Single-Purpose Eczema Power Ingredients

Managing eczema demands targeted ingredients that focus on specific aspects of barrier dysfunction, such as hydration, lipid restoration, or gentle cleansing, without exacerbating irritation. These single-purpose actives deliver precise relief for dryness, inflammation, or microbial imbalance, making them ideal for integration into gentle, fragrance-free routines. When paired with multi-functional ingredients, they create synergistic formulations that restore and maintain a healthy skin barrier, supporting long-term eczema control.

- **Bisabolol**: Chamomile extract that reduces redness and soothes sensitive or reactive skin without sensitization risk.
- **Ceramides**: Essential for restoring the compromised lipid barrier that characterizes eczema, preventing water loss while supporting overall barrier integrity.
- **Cholesterol and Fatty Acids**: Support optimal barrier lipid composition for proper barrier function and long-term skin health.
- **Cocamidopropyl Betaine**: Gentle cleansing agent that maintains barrier integrity during the cleansing process.
- **Hyaluronic Acid**: Attracts and retains essential moisture without irritation risk while actively supporting barrier repair processes. Use with care in dry, arid climates.
- **Glycerin**: Fundamental humectant that draws moisture to skin without irritation, suitable for daily use.
- **Oat Extract**: Similar to colloidal oatmeal but used in lighter formulations for soothing and barrier support.
- **Petrolatum**: Byproduct of crude oil refining that provides very heavy occlusive protection. Be sure to look for "USP grade" on labels in the US to ensure it is free of PAHs.
- **Postbiotics**: Provide benefits of healthy microbiome metabolites without introducing live bacteria that could cause reactions.
- **Prebiotics**: Support beneficial bacteria that help maintain healthy skin ecosystem and natural defense mechanisms.
- **Shea Butter**: An occlusive emollient that seals moisture, effective for eczema and widely used.
- **Sodium Cocoyl Isethionate**: Mild surfactant that cleanses effectively without stripping essential lipids from compromised skin.
- **Squalane**: Lightweight emollient that mimics natural skin oils, restoring suppleness without clogging pores or heaviness.
- **Urea (≤5%)**: Hydrates and gently exfoliates to address dryness and flakiness in eczema.

Building Your Eczema Strategy

Eczema management requires exceptional patience and consistency. Barrier repair and inflammation control typically take 4 to 6 weeks of dedicated care to show significant improvement. The key is using gentle, fragrance-free formulations designed specifically for sensitive, compromised skin.

Consistency is vital for managing eczema, far more than with other conditions. Maintaining a gentle routine even during clear periods helps prevent future flares.

Remember that eczema is a chronic condition that requires ongoing management, rather than a temporary problem with a permanent cure.

Focus on gentle, supportive care that works with your skin's natural healing processes rather than aggressive treatments that could trigger additional inflammation. Layer products from thinnest to thickest consistency and always prioritize barrier protection over targeted active ingredients that might cause irritation.

If gentle over-the-counter products haven't improved your eczema after 6 to 8 weeks of consistent use, consider consulting a licensed skincare professional who specializes in sensitive skin conditions. They can provide gentle professional treatments, deliver eczema-specific protocols, and guide you toward professional products designed for compromised barrier function.

If you have moderate to severe eczema that significantly impacts your quality of life, frequent skin infections, or eczema that doesn't respond to gentle care, consider consulting a dermatologist who can prescribe stronger treatments and develop a comprehensive management plan.

Psoriasis: An Autoimmune Inflammation Condition

Psoriasis is a complex autoimmune condition with three primary processes: dramatically accelerated cell turnover, chronic immune-mediated inflammation, and abnormal keratinocyte development. Psoriasis causes skin cells to multiply up to 10 times faster than normal, creating the characteristic thick, scaly plaques that define this skin condition.

Effective treatment must target excessive cell proliferation, reduce chronic inflammation, and help normalize the skin cell development process. The goal isn't just managing surface symptoms but addressing the underlying cellular chaos that drives plaque formation.

Lifestyle factors significantly influence psoriasis severity and flare frequency, making trigger identification foundational for comprehensive management. Stress acts as a major trigger by activating immune pathways that worsen inflammation. Infections, particularly streptococcal throat infections, can trigger guttate psoriasis flares in susceptible individuals. Some have found that a carnivore diet alleviates symptoms *(always consult with a medical professional before making dietary changes).*

Smoking and excessive alcohol consumption are associated with more severe disease progression. Certain medications, including lithium, beta-blockers, and antimalarials, may make symptoms worse. The Koebner phenomenon, where trauma to the skin from cuts, burns, or aggressive scratching triggers new psoriatic lesions at injury sites, makes gentle skincare particularly important. Weather changes, especially cold, dry conditions, often worsen symptoms by further compromising an already challenged skin barrier.

Moderate to severe psoriasis typically requires systemic treatments including biologics, oral medications, or phototherapy. Topical prescription treatments like vitamin D analogs, corticosteroids, and retinoids remain cornerstone therapies. Over-the-counter and professional skincare ingredients can provide important supportive therapy, helping manage the thick, scaly nature of psoriatic plaques while providing the intensive moisture support this condition demands. The goal is to work alongside medical treatments to improve comfort and appearance.

Multi-Function Psoriasis Ingredients

These multi-functional actives reduce scale buildup, soothe inflammation, help restore barrier integrity, and hydrate compromised skin, forming the cornerstone of effective psoriasis management. Found in high-quality, fragrance-free formulations, they deliver comprehensive relief, minimizing irritation and supporting long-term skin health for those with this challenging condition.

- **Centella Asiatica**: Promotes healing and reduces inflammation with proven efficacy for compromised skin.
- **Colloidal Oatmeal**: Gentle anti-inflammatory particularly beneficial for sensitive or newly affected areas.
- **Niacinamide**: Provides anti-inflammatory benefits with barrier support and soothing properties, helping calm the chronic inflammation characteristic of psoriasis.
- **Panthenol**: Hydrates, reduces inflammation, and promotes healing, a multi-functional staple for psoriasis-prone skin.
- **Salicylic Acid**: Essential keratolytic agent that dissolves thick, scaly buildup while promoting gentle exfoliation and providing anti-inflammatory benefits.
- **Urea (≤5%)**: Acts as both keratolytic and humectant, softening and removing scales while providing vital hydration through its moisture-binding properties.

Single-Purpose Power Ingredients

Managing psoriasis requires targeted ingredients that address scale removal, inflammation reduction, or moisture retention without increasing sensitivity. These single-purpose targeted actives deliver precise results, focusing on individual mechanisms such as exfoliation, soothing, or barrier protection. Ideal for integration into rich, fragrance-free formulations, they work synergistically with multi-functional ingredients to create tailored routines that alleviate symptoms and support the intensive care needs of psoriatic skin.

- **Allantoin**: Soothing and anti-inflammatory agent that promotes skin barrier healing and reduces irritation without any sensitization risk, ideal for reactive skin.
- **Alpha Hydroxy Acids**: Promote healthier cell turnover when used gently and consistently on thickened psoriatic skin.
- **Bisabolol**: Chamomile extract that reduces redness and soothes sensitive or reactive skin without sensitization risk.
- **Calendula Extract**: Traditional healing ingredient with anti-inflammatory properties and excellent safety profile.
- **Ceramides**: Essential for restoring barrier function in affected areas while providing the intensive moisturization that psoriatic skin desperately needs.

- **Coal Tar**: Traditional ingredient that reduces inflammation and slows excessive cell turnover, though less cosmetically elegant than modern alternatives.
- **Lactic Acid**: Provides gentle exfoliation with additional moisturizing properties, ideal for removing scale buildup.
- **Mahonia Aquifolium (Oregon Grape)**: Reduces inflammation and slows cell turnover, a natural alternative to coal tar for psoriasis.
- **Shea Butter**: Provides rich moisturization with natural anti-inflammatory properties for intensive barrier support.
- **Squalane**: Lightweight emollient that mimics natural skin oils, restoring suppleness without clogging pores or heaviness.
- **Glycerin**: Fundamental humectant that provides deep hydration essential for psoriatic skin's moisture needs. Use with caution in dry, arid environments.
- **Zinc Oxide**: Provides physical sun protection with mild anti-inflammatory and antimicrobial properties.

Building Your Psoriasis Strategy

Psoriasis, a chronic autoimmune condition, requires persistent, tailored care to manage scaling, redness, and discomfort, as complete clearance may not always be achievable with topical treatments alone. Consistent use of rich, fragrance-free, occlusive formulations designed for psoriasis or severely dry, thickened skin can significantly alleviate symptoms by penetrating psoriasis plaques and supporting barrier repair.

Initial improvements usually appear within 2–4 weeks, with optimal results developing over months of diligent care. Gentle application and layering products from thinnest to thickest helps prevent irritation and the Koebner phenomenon, where new lesions form from skin trauma.

If over-the-counter treatments yield insufficient relief after 6–8 weeks, if you haven't already, consult a dermatologist or licensed skincare professional for specialized protocols, professional exfoliation, or prescription options to address severe or stubborn psoriatic skin effectively.

LABEL LITERACY: KNOW WHAT YOU'RE REALLY BUYING

Understanding skincare product labels is a little like learning a new language, and that can feel a little overwhelming at first. In this chapter, you'll learn how to decode the regulatory frameworks that govern ingredient labeling, understand why the order of ingredients matters more than you might think.

You'll learn to recognize marketing claims that don't match the actual formulation, understand some hidden ingredients that companies don't have to disclose, identify common irritants that could be sabotaging your skin goals.

By the end of this chapter, you'll have the ability to better evaluate any skincare product based on its actual merit rather than its marketing appeal. You'll know how to spot "fairy dusting," understand what some of those intimidating scientific names mean, and make better informed decisions that align with your skin's needs and your budget.

What the Label *Really* Means

Skincare product labels aren't just random lists-they follow strict rules that vary by country to ensure safety and transparency. These regulations dictate which ingredients are allowed and the ingredients are listed on skincare ingredient labels. Understanding these rules can guide you in choosing the products that work best for your skin, no matter where you are.

How Ingredients Are Listed

In most countries, ingredients are listed in order of how much is in the product, from highest to lowest concentration. Here's how it works in key regions:

- **United States**: The FDA requires ingredients to be listed in descending order for anything above 1% of the product. Below 1%, they can be listed in any order. Only 11 ingredients are banned, so you'll see a wide range of actives in U.S. products.
- **European Union**: The EU's strict Cosmetic Products Regulation lists ingredients in descending order and bans over 1,600 substances for safety. It uses the INCI (International Nomenclature of Cosmetic Ingredients) system for standardized names and requires fragrance allergens (like linalool or limonene) to be listed if they're above 0.01% in rinse-off products or 0.001% in leave-on products, which is helpful for sensitive skin.
- **Canada**: Follows rules similar to the EU under the Food and Drugs Act, with a focus on safety and clear labeling, using the INCI system.
- **Japan**: Uses Japan Cosmetic Ingredients (JCI) standards, with strict safety checks and standardized naming for clarity.
- **South Korea**: Combines U.S. and EU approaches under the Ministry of Food and Drug Safety, emphasizing safety and detailed ingredient lists.
- **Southeast Asia**: The ASEAN Cosmetic Directive aligns rules across countries like **Thailand and Singapore**, similar to the EU's focus on safety.
- **China**: Recently updated rules to reduce animal testing, with ingredient lists following descending order and safety-focused regulations.

INCI: The Global Ingredient Label Language

If you've read skincare labels, you've seen water listed as *Aqua* and shea butter listed as *Butyrospermum Parkii*. These are the official names that are defined in the ***International Nomenclature of Cosmetic Ingredients*** (INCI). The globally adopted INCI is a standardized system for naming ingredients used in cosmetics, skincare, and other personal care products for ensuring consistency and transparency in labeling.

INCI names on skincare ingredient labels are mandatory in regions like the European Union and are widely used in the U.S., Canada, and other markets. This worldwide standard helps make skincare and cosmetic labels clear and consistent. Learning to recognize INCI names helps you understand exactly what's in your skincare products, no matter the brand's fancy marketing terms, so you can choose what's best for your skin.

What Are INCI Names?

INCI names are standardized to identify ingredients the same way across countries, avoiding confusion from different common names. Names are assigned based on the ingredient's chemical composition, botanical origin, or function. For example:

- **Plant ingredients**: Use Latin botanical names plus the plant part, like *Chamomilla Recutita Flower Extract* for chamomile flower extract.
- **Chemical ingredients**: Use scientific names, like Tocopherol for vitamin E or Ascorbic Acid for vitamin C.
- **Colors**: Listed as "CI" with a five-digit number, like CI 77891 for titanium dioxide (a white pigment often in sunscreens).
- **Synthetic ingredients**: Have complex chemical names, like Dimethicone (a silicone for smoothness) or Phenoxyethanol (a preservative).

Why INCI Names Matter

Knowing INCI names lets you spot key ingredients and understand their role, even if brands use creative names. For example:

- **Multiple roles**: Some ingredients do more than one job. Cetyl Alcohol can smooth skin (as an emollient) or hold products together (as an emulsifier). Its place on the label (higher for more, lower for less) shows its main purpose.
- **Avoid confusion**: Don't be fooled by "alcohol-free" claims. Fatty alcohols like Cetyl Alcohol or Stearyl Alcohol hydrate skin, unlike drying alcohols like Alcohol Denat.
- **Spot ingredient families**: Look for patterns, like silicones ending in "-cone" or "-siloxane" (e.g., Dimethicone), which add smoothness, or sulfates like Sodium Lauryl Sulfate, which can be harsh for sensitive skin.

By getting familiar with common INCI names, you can quickly compare products, spot ingredients that work for you (like niacinamide for oil control), and avoid ones that don't (like fragrances for sensitive skin). This makes picking the right skincare easier and helps you build a routine that keeps your skin healthy and glowing.

The INCI Decoder & Other Apps

To make decoding labels easier, apps like INCI Decoder, CosDNA, and EWG's Skin Deep provide searchable databases where you can enter a product name or ingredient name in INCI name or in common form. The ingredient entries provide the ingredient's role (e.g., solvent, humectant, preservative) and sometimes a bit about how it functions. These tools are important for helping you identify target actives, potential allergens, irritants, comedogens, and other ingredients that the skincare label reveals.

Skincare Label Ingredient Order Tells a Story

No matter where you shop, the first 5 to 7 ingredients on a label usually make up 80% to 90% of the product, so they're generally the ones doing most of the work. Ingredients listed toward the end, such as targeted actives, peptides, or advanced regenerative ingredients, can have tiny effective concentrations (0.1% or less).

Common end-of-list ingredients also include fragrances and preservatives, which may matter for sensitive skin. Regional regulatory differences might mean that a product's formula or label changes depending on where it's sold, so check labels carefully if you're buying globally.

Ingredient Concentration Matters

Ingredients are ordered by concentration, so any ingredient's place gives you an idea of its concentration in the product. If a product claims to be "packed with hyaluronic acid" but lists it on the label after Phenoxyethanol (a preservative typically at 1% or less), there's likely very little hyaluronic acid, so it may not hydrate as promised. Knowing this helps you decide if a product's price matches its quality.

Some ingredients need higher concentrations to work their best and should appear early on the label if they're the star players:

- **Niacinamide** (4–5%): Controls oil and brightens skin.
- **Vitamin C** (10–20%): Fades dark spots and fights aging.
- **Retinol** (0.1–1%): Smooths wrinkles and boosts cell turnover.
- **Salicylic Acid** (1–2%): Clears pores for acne-prone skin.

Ingredients like water (*Aqua*) often make up 60% to 80% of the product formulation, so they're usually listed first. Oils or ingredients like shea butter (*Butyrospermum Parkii*) in the first five ingredients suggest a rich, creamy texture, great for dry skin. Products with humectants like glycerin or hyaluronic acid as the first ingredients on the list with few emollients or oils will feel lighter and suit oily or combination skin. Where ingredients fall in the ingredient list order helps gauge concentrations and formulations so you predict how a product will perform and feel before buying.

Don't fall for marketing hype about "trendy" ingredients like "rare orchid extract", especially if they're at the end of the list. Their concentration is likely too low (0.1% or less) to make a big difference. In general, focus on the first 5 to 7 ingredients to understand a product's main benefits and effectiveness.

Some skincare ingredients work their wonders in tiny amounts (less than 1%) because they target specific skin processes, like boosting collagen or calming inflammation.

Higher concentrations of these potent ingredients can cause irritation and other issues. Common examples include:

- **Peptides:** 0.01–0.1% for smoothing wrinkles and firming skin.
- **Bakuchiol:** 0.5–1% for anti-aging, a gentler retinol alternative.
- **Azelaic Acid:** 0.1–1% for reducing redness and acne.
- **Tretinoin:** 0.025–0.1% for powerful wrinkle and acne treatment.
- **Growth Factors:** 0.0001–0.01% for skin repair and renewal.

By reviewing ingredient order and understanding concentrations, you can choose products that match your skin's needs, avoid overhyped claims, and build a routine that keeps your skin healthy and glowing.

Formulation Red Flags

Even the best skincare products can include ingredients that irritate certain skin types. Knowing common culprits and warning signs helps you choose products that keep your skin healthy and comfortable, especially if you have sensitive or reactive skin. This section highlights ingredients to watch for and tips to avoid irritation, making it easier to build a routine that works for you.

Fragrance: The #1 Troublemaker

Fragrance (listed as *Fragrance, Parfum, Natural Fragrance,* or *Aroma*) that can hide dozens of chemicals is the most common cause of skin irritation and allergies in skincare. These terms can hide dozens of chemicals, as brands don't have to list them due to trade secret rules. Even "natural" fragrances, like essential oils from lavender, bergamot, or lemon, can irritate or make skin sensitive to sunlight, causing dark spots.

Products labeled "unscented" might still have masking fragrances to cover other scents, so always look for "fragrance-free" if your skin is sensitive. In the EU, labels must list 26 common fragrance allergens (especially Linalool and Limonene) if they're above low levels, but in the U.S., these can stay hidden. For sensitive skin, fragrance-free products are the safest bet.

Alcohol: Not All Are Bad

Not all alcohols harm your skin - it depends on the type. Simple alcohols like *Alcohol Denat.* or *Isopropyl Alcohol* can dry out or irritate skin, especially in high concentrations, but they're okay for oily skin if balanced with moisturizing ingredients. Check their position on the label. The higher they are in the list means more drying potential.

Fatty alcohols like *Cetyl Alcohol, Stearyl Alcohol,* and *Cetearyl Alcohol* are different – these alcohols hydrate and smooth skin, making them great for dry skin in creams or lotions. Don't avoid a product just because it says "alcohol." Understanding how much of which kind helps you know if it will be helpful or harmful for your skin.

Sulfates: Harsh Cleansers

Sulfates like *Sodium Lauryl Sulfate (SLS)* or *Sodium Laureth Sulfate (SLES)* are strong cleansers in face washes that can strip natural oils, irritate sensitive or dry skin, or worsen conditions like eczema. *SLES* is slightly gentler than *SLS* but can still be drying.

Look for milder alternatives like *Cocamidopropyl Betaine* or *Sodium Cocoyl Isethionate* (from coconut) that clean well without harming your skin's barrier. If your skin feels tight or irritated after cleansing, try a sulfate-free option for better comfort.

Preservatives: Necessary but Tricky
Preservatives keep products safe by preventing germs, but some can irritate sensitive skin. *Methylisothiazolinone (MI)* and *Methylchloroisothiazolinone (MCI)* are common culprits, often causing redness or rashes, and are restricted in Europe for leave-on products. Formaldehyde-releasing preservatives like *DMDM Hydantoin* and *Diazolidinyl Urea* cause reactions in about 8% to 9% of people sensitive to formaldehyde. Safer options include *Phenoxyethanol* or *Benzoic Acid*, often paired with helpers like *Disodium EDTA* to use less preservative. Always patch-test products (see previous section *Patch Testing & Tracking Results*) with these ingredients if you have sensitive skin.

Essential Oils: Natural but Not Always Gentle
Despite being natural, essential oils like lavender, peppermint, or citrus like bergamot and lemon can irritate sensitive skin or increase sun sensitivity, leading to burns or dark spots. For example, citrus oils have compounds (furanocoumarins) that react with sunlight, and peppermint's menthol can sting.

Even lavender, often seen as gentle, contains allergens like *Linalool*. If you have sensitive or reactive skin, stick to fragrance-free products, especially for your face, where skin is thinner and more prone to reactions.

Hidden Irritants: Sensations Aren't Always Good
Products that "tingle," "heat up," or feel "active" might sound effective but often signal irritation. Ingredients like menthol, camphor, or high-strength glycolic acid can cause stinging or redness, which may harm your skin over time. Physical scrubs with rough particles such as walnut shells or apricot kernels can create tiny tears, weakening your skin's barrier. Even witch hazel can be harsh due to its drying compounds. Chronic irritation from these can make skin more sensitive, so choose gentle products that don't sting or tingle for healthier results.

"Hidden" Ingredients: What Labels Don't Show
Some ingredients or by-products aren't listed on labels but can cause issues, especially for sensitive skin:

- **1,4-Dioxane**: This trace contaminant can form in ethoxylated ingredients, meaning those with PEG, Polyethylene, and -**eth**-, or -**oxynol** suffixes, like *Sodium Lau**reth** Sulfate* or *Non**oxynol**-9*. It's not listed because it's a by-product, not an added ingredient, but it may pose concerns for sensitive skin.

- **Vague Terms**: Label ingredients like "*proprietary blends*," "*natural fragrance*," or "*coconut-derived surfactant*" can mask specific chemicals that might irritate. Even "*flavor*" in lip products can hide allergens.

- **Other By-Products:** Unlisted traces – such as formaldehyde (released from some preservatives) - or other manufacturing residues may also cause reactions.

To avoid surprises, choose brands that fully disclose ingredients or opt for fragrance-free products, which are less likely to include hidden irritants like phthalates or 1,4-Dioxane. Patch-test new products (see previous section *Patch Testing & Tracking Results*) to ensure they suit your skin.

Controversial Ingredients: Know the Debate
Some ingredients are safe by regulation, but spark debate due to safety or environmental concerns. Here's what to know to make better choices that fit your needs:

- **Parabens** (*Methylparaben, Propylparaben*): Used as preservatives, these raised concerns after a 2004 study found them in breast tumors, hinting at hormone disruption. The FDA and EU say they're safe at low levels (0.01–0.3%), but some avoid them due to distrust in long-term safety. Many brands now use alternatives like *Phenoxyethanol.*

- **Chemical Sunscreens** *(Oxybenzone, Octinoxate):* These protect against UV but are criticized for coral reef damage (banned in Hawaii) and absorption into the bloodstream. Mineral options like Zinc Oxide or Titanium Dioxide are popular alternatives, though they may feel heavier.

- **Silicones** ***(Dimethicone, Cyclopentasiloxane)***: These smooth the skin but face scrutiny for environmental buildup enough for the EU to restrict some cyclic silicones They're deemed safe for skin but may not suit those prioritizing eco-friendly products.

- **Sulfates (*SLS, SLES*)**: Beyond irritation, concerns linger about manufacturing by-products such as *Ethylene Oxide* that can cause long-term health issues, particularly cancer, from chronic low-level inhalation. Although they're deemed safe by regulators, gentler cleansers like *Coco-Glucoside* are preferred by many.

Weighing these controversies means balancing science, personal values, and skin needs. Look at independent research, consider safer alternatives, and test products to ensure they work for you without irritation.

Skincare Marketing Hype Decoder

Skincare marketing sells feelings and "transformation" stories, not formulas. Claims often muddy the boundary between science and marketing hype. This section helps you spot misleading claims so you can better choose products that meet your skin's needs.

Meaningless Buzzwords & Claims
Many terms sound impressive but have no standard definition or testing to support them. Here's what to watch for:

- **Anti-Aging** – No scientific definition and no standard for what counts as "anti-aging" benefits; it's a catch-all, overused umbrella term.
- **Award-Winning**: Many beauty awards are pay-to-play or editorial promotions. An "award" doesn't equal third-party verification or scientific credibility.
- **Backed by Consumer Trials**: These are usually just surveys with percentages; not the same as controlled scientific studies.
- **Clinically Proven / Clinically Tested**: Meaningless without details (sample size, duration, funding, dropout rates, endpoints, peer review). Tiny subjective studies with poor scientific method use are often behind this phrase.
- **Cruelty-Free**: No single regulatory standard; any brand can claim it. Look for strict certifications (e.g., Leaping Bunny, PETA's Beauty Without Bunnies).
- **Dermatologist** Recommended / Tested: Can mean paid endorsements or minimal "testing." One dermatologist, one person, one time still qualifies.
- **Editor's Choice**: Free product, ad revenue, or other compensation can compromise objectivity.
- **Fragrance-Free**: May still include fragrant components or masking agents; always read the label.
- **Hypoallergenic**: No regulated definition or required testing; reactions are still possible.
- **Medical Grade / Clinical Grade / Professional Strength**: Marketing phrases without recognized skincare regulatory meaning; not a guarantee of stronger or better results.
- **Microbiome-Friendly**: Vague and unregulated; doesn't guarantee outcomes.
- **Natural**: "Natural" ≠ better or safer (poison ivy is natural).
- **Non-Comedogenic**: No universal standard or test; ingredients still matter.
- **Non-Toxic / Toxin-Free**: Vague fear-based claim. The dose makes the poison.
- **Organic**: Standards vary by country; the product may be only partly organic; not a results guarantee.
- **Paraben-Free / Sulfate-Free**: Often used to stoke fear rather than address real, ingredient-specific safety.
- **Preservative-Free**: Sounds "pure," but water-containing products without preservatives risk bacteria/mold growth.
- **Suitable for Sensitive Skin**: Suggests gentle but isn't regulated and won't suit every sensitivity.
- **Vegan**: No animal-derived ingredients, but doesn't guarantee efficacy, safety, or cruelty-free status.

Pseudo-Scientific Claims

Brands often use science-sounding terms to seem cutting-edge, but many are empty without proof:

- **Adaptogens**: Claimed to calm skin, but there's little evidence they work topically.
- **Blue Light Protection**: Limited proof that screen light harms skin or that products block it.
- **Chemical-Free**: Utter nonsense - everything in the universe - all matter, both living and non-living, is made of chemicals.

- **Clinically Proven**: Sounds legit but might come from a small, brand-run study. Ask: How many people were in the study? What was measured? Was it independent?
- **Cosmeceutical**: A made-up word implying drug-like benefits without strict testing.
- **DNA Repair**: Usually, an exaggerated claim that may be legitimate for newer advanced regenerative biotechnology, particularly for professional products designed for in-office use. As a rule, topical products can't repair DNA, though some actives like peptides, growth factors, and stem cell derivatives can support improved skin function.
- **Patented Formula**: Al this means is that the formula is unique, not that it is effective.
- **Pharmaceutical Grade**: This is a regulated term. When Pharmaceutical Grade is on a label, it means that the product meets stringent purity standards set by regulatory bodies like the U.S. Pharmacopeia (USP) and European Pharmacopoeia (EP). Note that purity doesn't equal effectiveness.
- **Proprietary Complex**: Hides ingredient amounts so you can't tell if actives like Hyaluronic Acid are in concentrations high enough to work.

Magical Promises That Don't Deliver

Some claims sound life-changing but are unrealistic:

- **Erases Wrinkles/Instant Facelift**: Wrinkles form over time due to collagen loss, repetitive facial movements, and environmental damage like UV exposure. No topical product can erase them completely or mimic the structural lift of a surgical facelift. Some formulas may temporarily smooth or tighten the skin by using film-forming agents, silicones, or light-diffusing particles, but these effects are cosmetic and short-lived.
- **Pore-Reducing**: Pore size can't physically be reduced or shrunk. Skincare ingredients that reduce oil production, clear debris, and/or improve skin elasticity can make them look smaller though.
- **Reverses Aging**: Aging is a complex, multifactorial process involving genetics, environmental exposure (like UV damage), and internal cellular changes. There are ingredients like retinoids, peptides, and antioxidants in topical skincare products that can improve the appearance of fine lines, boost hydration, and support skin renewal, however at best they just slow or soften visible signs of aging. To date, aging's underlying biological clock can be slowed, but nothing can reverse it.
- **Instant Results**: Temporary plumping effects from silicones, caffeine, or peppermint for lips aren't lasting results. True skin changes involve biological processes like cell turnover, collagen synthesis, and barrier repair, which take time. Instant results can also mask irritation with tingling or tightness sometimes being mistaken for effectiveness.
- **Unrealistic Before & After Photos**: Images, especially those that look like us, are powerful marketing tools. They're often manipulated to exaggerate results with lighting, angles, facial expressions, makeup, photo retouching, and can even show results from other treatments like Botox, not the skincare product. Some brands also use different models or timeframes that don't reflect typical results. When photos promise dramatic transformations in days, it's a red flag.

Look instead for clinical trial data, ingredient transparency, and realistic skin condition improvement timelines.

Tricky Numbers & Statistics

Every marketer has read 1954's *How to Lie with Statistics* by Darrell Huff, the primer for using statistics to deceive. Marketers use statistics to create a sense of scientific credibility of claims that may or may not be so truthful or meaningful.

- **Study Results**: Claims like "90% saw smoother skin" or "4× more hydrating" sound great but can mislead: They might come from tiny studies with as few as 5 people or vague, subjective survey questions like "My skin feels smoother. Yes or No."

- ***X* More**: "10× more active ingredients" sound impressive. Which "active ingredients"? Is 10x more based on concentration, number of ingredients, or perceived potency? Brands may count any ingredient with a skin-related function as "active," even if it's present in trace amounts or has minimal effect. Then there's the fact that more isn't always better: piling on actives can increase the risk of irritation, especially for sensitive skin.

- **Cherry Picking**: Brands can cherry-pick statistical results by selectively highlighting only the most favorable outcomes from studies and ignoring what's less flattering. For example, a brand might claim "90% saw brighter skin," but omit that only 10% saw increased wrinkles, or that the study excluded participants with sensitive skin.

It's so important to look closely at the source of glowing statistical claims, especially if the numbers or statistics are a factor in your product decision. Always review study details for size, methods, and independence before trusting the numbers. If a study looks too technical to understand, you can upload the document or provide the URL to any AI and ask it to summarize the study and findings in lay language.

Luxury Pricing and Marketing Gimmicks

These tactics create artificial scarcity or "worth" regardless of formulation quality. Heavy frosted glass, metallic lids, and story-driven branding signal luxury pricing. Price ≠ performance. There are luxury brands that are effective due to inexpensive active ingredients like retinols and niacinamide that with no active ingredients that justify the luxury price. To protect your wallet, watch for:

- **Doctor's Brand**: Uses medical authority to justify cost, but ingredients like Vitamin C are the same as in less expensive products.
- **Influencer/Celebrity Collaborations**: Boosts hype, not quality or effectiveness.
- **Emotional Storytelling**: Sells a "self-care" vibe, not better a formulation.
- **Korean/French/Swiss Skincare**: Geographic labels sound exotic but don't guarantee results.
- **Limited Edition**: Creates urgency (FOMO) but says nothing about effectiveness.

How to Shop Smarter

Shopping smarter means simply focusing on facts:

- **Read Ingredient Labels:** Look for proven ingredients appearing in expected order on the ingredient list.
- **Assess Concentrations:** Avoid "fairy dusting" where actives are too low, like a "Retinol Night Cream" where *retinol/retinyl palmitate* is listed after *fragrance.*
- **Be Patient:** Real results typically take weeks.
- **Research Claims:** Seek out independent studies, not brand promises.
- **Patch-Test:** The bottom line is how well a product does or does not work to provide the support your skin needs. Always patch test new products, especially for sensitive skin (see previous section *Patch Testing & Tracking Results*).

By focusing on ingredients and evidence, you can pick products that deliver real results, saving time and money while keeping your skin healthy and glowing.

About Professional Products

There are two main types of skincare products; Over-The-Counter (OTC) products sold at stores and "professional" products available through dermatologists, medical spas, or licensed aestheticians. This section explains what makes professional products different, when they're worth the investment, and how to navigate potential pitfalls to build a routine that works for your skin.

What Makes a Product "Professional"?

The term "professional" isn't regulated, but these products stand out in three practical ways:

- **Higher Active Ingredient Concentration:** Professional products often have stronger concentrations of active ingredients than OTC versions. For example, a retail vitamin C serum might have 5–15% L-Ascorbic Acid, while professional ones may have 15–20% with optimized pH for better results. These higher doses can tackle tough skin issues but may irritate if not used correctly, so professional guidance is key.
- **Advanced Research:** Professional brands invest heavily in clinical testing, stable formulations, and proprietary delivery systems that ensure ingredients like peptides and growth factors penetrate effectively and work without causing irritation.
- **Exclusive Distribution:** True professional products are sold through dermatologists, aestheticians, or medical spas, not mass retailers. This ensures expert advice to prevent misuse of potent formulas and maintains the brand's premium image.

Retail Sub-Brands: Know the Difference

Many professional brands now sell "retail" versions online or in stores, which can be confusing. These sub-brands often have:

- Lower active ingredient concentrations
- Simpler formulas without advanced delivery systems

- Consumer-friendly packaging but the same brand name.

To spot differences, compare ingredient lists and concentrations between professional and retail versions. Some sub-brands keep decent potency, while others dilute key actives significantly.

When to Go Professional

Not everyone needs professional-grade skincare products, however when there are persistent skin concerns, post-procedure support is needed, or there complex or overlapping skin conditions, professional skincare products are worth the investment. Professional skincare lines include specialized products with concentrated ingredients for chronic acne, rosacea, sensitive, stubborn pigment, and other skin concerns.

Aging skin usually needs more concentrated, better-delivered actives to overcome slower renewal and structural loss. Today's most advanced regenerative biotechnology ingredients - growth factors, exosomes, and next-generation peptide systems- are only available in professional skincare products, some that must be applied in-office. Professional skincare products for post-procedure support accelerates healing and help amply results following professional treatments such as chemical peels, laser skin resurfacing, microchanneling, or microneedling.

The Role of Skin Professionals

Dermatologists and licensed skincare professionals provide more than just product recommendations - they bring expertise to transform your skincare:

- **Accurate Diagnosis**: They can identify underlying issues that aren't immediately visible to target root causes, not just symptoms.
- **Smart Regimen Design**: They pair ingredients that work together and avoid risky combinations. They also create gradual plans to build tolerance, ensuring better results without irritation.
- **Treatment Integration**: Professionals combine in-office treatments with home products to maintain results. They troubleshoot issues like purging, saving you time and money.

Navigating Conflicts of Interest

Professional products can be pricey, and sales are a big revenue source for the practices that sell them. Practitioners typically earn commissions and often must meet organizational sales quotas, which can create bias. To make informed choices:

- **Ask Why**: Request clear, ingredient-based reasons for recommendations (e.g., why vitamin C at 20% is better for your dark spots).
- **Compare Options**: Search for products with similar ingredients are available at better prices through other brands or OTC channels.
- **Get a Second Opinion**: If pushed toward an expensive regimen or single brand, consult another professional for balance.
- **Watch for Red Flags**: Be cautious if a practitioner:
    - Refuses to discuss alternative ingredients or insist that only their products work.

- Pressures you to buy large quantities or "exclusive" deals immediately.
- Cites vague brand claims without explaining benefits.
- Discourages researching alternatives or seeking other opinions.

How to Choose Wisely

To get the most from professional skincare:

- **Compare Ingredient Labels**: Compare professional and retail products for active concentrations.
- **Seek Expert Guidance**: Use dermatologists or licensed skincare professionals for complex issues or post-procedure care but ask for evidence-based advice.
- **Start Slow**: Patch test potent products on a small area to avoid irritation, especially if you have sensitive skin.
- **Balance Cost and Value**: Professional products can be worth it for stubborn concerns, but don't overpay for diluted retail sub-brands.

By understanding what makes professional products special and when they're needed, you can make smart choices that keep your skin healthy, glowing, and thriving.

PART II: THE SKINCARE INGREDIENT REFERENCE

Part I of this book builds the foundation, explaining how your skin functions, how to assess its needs, how to interpret marketing claims, and how to understand ingredient labels so you can make confident, informed decisions. The next step is a deeper dive into how individual skincare ingredients work. The challenge is scale. Thousands of ingredients exist, many with overlapping roles and functions, some with multiple aliases, and some with inconsistent claims.

To keep the reference practical and user-friendly, **Part II - The Skincare Ingredient Reference** is organized by ingredient function. This allows you to quickly find the categories most relevant to your skin concerns. This guide is designed to be your companion for apps like the INCI Decoder and CosDNA that provide ingredient high level ingredient categories but lack details into how ingredients function and how they work or don't work well with other ingredients.

Each section begins with a short introduction that explains how that ingredient category contributes to overall skin health and product performance. Ingredient details in each section focus on the most clinically relevant and commonly used ingredients, the ones you're most likely to find on product labels, in marketing materials, or in professional treatments.

Every ingredient entry follows a consistent format so you can easily scan for what matters most. Skincare ingredients are listed alphabetically using their INCI name (see section *INCI: The Global Ingredient Label Language* in PART I).

Ingredient details are provided in the format below so that you have insight into the ingredient's function, optimal concentration range, synergistic pairings, and known limitations. This approach is designed to help you interpret labels with the same level of understanding as a trained formulator or skincare professional.

[INCI Name] (any common or brand name) (ingredient role, effective concentrations that tell you where the ingredient should be on the label)

Best For: Indicates suitable skin types or product categories where the ingredient performs optimally.
What It Does: A brief, function-focused description outlining the ingredient's cleansing, toning, or rejuvenating capabilities.
Effectiveness: An assessment of the ingredient's performance, often noting mildness, daily-use compatibility, and skin tolerance.
Pair With: Suggested compatible ingredients that enhance function or support skin barrier health.
Avoid Using With: Highlights ingredient or formulation types that may compromise performance or stability.
Notes: (Optional) Any additional considerations, such as sourcing, regulatory context, or formulation nuances that are important for you to know.

Because many actives are multifunctional (for example, niacinamide and ascorbic acid), some ingredients appear in more than one section. This intentional duplication allows you to see how an ingredient performs in each functional context without excessive cross-referencing.

Skincare ingredients are evolving. Skin health has become recognized as a biomarker of longevity, a visible indicator of systemic health and biological aging. This new status has led to a surge in regenerative biotechnology investment. There are new classes of peptides, exosomes, and bioengineered actives in the ingredient development pipeline. Some are in development, others undergoing trials, and others are waiting for regulatory approval. The *Advanced Regenerative Biotechnology* section of this reference includes only those ingredients already in commercial use or with established safety data at the time of publication.

Part II provides you with a working reference companion to the INCI database - one that bridges the gap between technical words and practical understanding. Whether you're evaluating a label, verifying a marketing claim, or comparing options for a specific concern, this section helps you:

- Recognize ingredient functions and limitations
- Understand synergistic ingredient pairings
- Make evidence-based, confident skincare product choices

This isn't a just skincare ingredient reference; it's a decision tool. This is your guide to reading between the lines of any ingredient label and understanding exactly what you're putting on your skin.

CLEANSING & PURIFYING

The foundation of any effective skincare routine begins with proper cleansing. Cleansing agents remove accumulated debris, excess sebum, environmental pollutants, makeup, and sunscreen that would otherwise clog pores and create a barrier between your skin and active ingredients. Proper cleansing is essential for maintaining healthy skin function and preventing issues like breakouts, dullness, and irritation, while also ensuring that the beneficial ingredients in your serums and moisturizers can reach and benefit your skin.

Toners refresh and soothe skin, help remove residual impurities and excess oil, restore your skin's pH balance, and prepare it for better absorption of subsequent skincare products. Modern toners have evolved far beyond the harsh, alcohol-based astringents of the past to include hydrating essences, exfoliating acid treatments, and barrier-supporting formulations that can address specific skin concerns while providing an essential bridge between cleansing and moisturizing.

Understanding the specific ingredients that make cleansing products helps you choose formulations that match your skin's needs and tolerance levels. The following sections break down the key categories of cleansing ingredients, from the surfactants that lift away impurities to the specialized agents that ensure different product phases work together harmoniously, giving you the knowledge to select cleansers that clean effectively without compromising your skin barrier.

Smart Cleansing: Surfactants 101

The foundation of any effective skincare routine begins with proper cleansing. Cleansing agents remove accumulated debris, excess sebum, environmental pollutants, makeup, and sunscreen -all of which can clog pores and create a barrier between the skin and active ingredients. Proper cleansing is essential for maintaining healthy skin function and preventing issues such as breakouts, dullness, and irritation, while also ensuring that the beneficial ingredients in serums and moisturizers can effectively penetrate the skin.

Surfactants are key ingredients in cleansers; they help remove dirt, oil, and makeup while producing foam for a satisfying cleanse. Not all surfactants are the same - some are gentle enough for sensitive skin, while others are formulated for oily or acne-prone skin. This section explains common surfactants, their benefits, and how to choose the right ones for your skin, helping you build a routine that keeps your skin clean, balanced, and healthy.

Mild, Plant-Derived Surfactants

These gentle surfactants are ideal for sensitive, dry, or eco-conscious skin types, providing effective cleansing without stripping the skin's natural barrier.

C10-16 Alkyl Glucoside (Non-Ionic, 2–6%)

Best For: All skin types.
What It Does: A plant-derived blend that gently cleanses with stable foam, versatile for various skin needs.
Effectiveness: Great for daily AM/PM cleansing, maintaining skin comfort.
Pair With: Glycerin, Betaine (hydration); botanical extracts (soothing).
Avoid Using With: Strong acids/bases or heavy clays, which may reduce performance.

Caprylyl/Capryl Glucoside (Non-Ionic, 1–4%)

Best For: Sensitive skin, baby products, natural formulas.
What It Does: A sugar-based surfactant that gently cleanses, solubilizes oils, and stabilizes emulsions.
Effectiveness: Ultra-mild for daily use, perfect for sensitive skin.
Pair With: Natural oils, Panthenol, Ceramides (barrier support).
Avoid Using With: Strong preservatives or high pH formulas.

Coco Glucoside (Non-Ionic, 0.5–3%)

Best For: Sensitive skin, eco-friendly products.
What It Does: Made from coconut and fruit sugars, it cleanses lightly with soft foam, preserving skin's pH.

Effectiveness: Gentle for daily use, eco-friendly, and well-tolerated.
Pair With: Decyl Glucoside, Glycerin, Aloe Vera.
Avoid Using With: Strong acids or hard water (less foam).

Decyl Glucoside (Non-Ionic, 1–4%)

Best For: Sensitive, dry, or compromised skin.
What It Does: A coconut- and sugar-based cleanser that's ultra-gentle, ideal for delicate skin.
Effectiveness: Cleanses without irritation, best for sensitive or baby skin, but may be too mild for heavy makeup.
Pair With: Panthenol, Ceramides, Glycerin.
Avoid Using With: Strong acids or heavy occlusives (reduces foam).

Lauryl Glucoside (Non-Ionic, 3–8%)

Best For: Sensitive skin, eco-conscious formulas.
What It Does: A coconut- and corn-derived cleanser with mild foam, biodegradable and gentle.
Effectiveness: Good for daily cleansing with minimal irritation.
Pair With: Coco Glucoside, Hyaluronic Acid, Chamomile.
Avoid Using With: High salts or very acidic formulas.

Amino Acid-Based Surfactants

These ultra-mild surfactants, derived from coconut and amino acids, are perfect for sensitive or reactive skin, providing gentle cleansing with creamy foam.

Disodium Cocoyl Glutamate (2–8%)

Best For: Sensitive, eczema-prone skin.
What It Does: Cleanses gently with soft foam, maintaining pH and moisture balance.
Effectiveness: Ideal for twice-daily use, especially in clear formulas.
Pair With: Sodium Cocoyl Glutamate, Glycerin, Ceramides.
Avoid Using With: Harsh sulfates (SLS) or drying alcohols.
Notes: Common in Japanese/Korean gentle cleansers.

Potassium Cocoyl Glycinate (2–10%)

Best For: Sensitive skin, baby care, atopic dermatitis.
What It Does: Provides ultra-gentle cleansing with creamy foam, preserving skin's pH.
Effectiveness: Mild enough for frequent use without dryness.
Pair With: Sodium Cocoyl Glycinate, Allantoin, Sodium PCA.
Avoid Using With: Harsh sulfates or strong acids/bases.

Sodium Cocoyl Glutamate (3–10%)

Best For: Sensitive, pH-balanced cleansing.

What It Does: Gently cleanses with soft foam, respecting skin's barrier.
Effectiveness: Great for daily use on delicate skin.
Pair With: Disodium Cocoyl Glutamate, Betaine, Ceramides.
Avoid Using With: Harsh sulfates or drying alcohols.

Sodium Cocoyl Glycinate (2–10%)

Best For: Sensitive, rosacea-prone skin.
What It Does: Ultra-gentle with creamy foam, cleanses without irritation.
Effectiveness: Ideal for frequent use, dermatologist recommended.
Pair With: Potassium Cocoyl Glycinate, Niacinamide, Hyaluronic Acid.
Avoid Using With: Harsh surfactants or low pH (<4).

Mild Anionic Surfactants

These provide a balance of effective cleansing and gentleness, often used as sulfate-free alternatives with rich foam.

Disodium Laureth Sulfosuccinate (5–15%)

Best For: All skin types, sulfate-free formulas.
What It Does: A mild alternative to SLES, creates stable foam without stripping oils.
Effectiveness: Great for daily cleansing with a luxurious lather.
Pair With: Cocamidopropyl Betaine, Glycerin, Panthenol.
Avoid Using With: High cationic surfactants (may cause precipitation).
Notes: Popular in sulfate-free shampoos and cleansers.

Sodium Lauroyl Methyl Isethionate (3–10%)

Best For: Sensitive skin, sulfate-free cleansers.
What It Does: Creates creamy foam, cleanses gently without dryness.
Effectiveness: Luxurious for daily use, keeps skin soft.
Pair With: Panthenol, Aloe Vera, Glycerin.
Avoid Using With: Highly acidic actives (Vitamin C).

Sodium Methyl Cocoyl Taurate (3–12%)

Best For: All skin types, "clean beauty" formulas.
What It Does: A coconut- and taurine-derived surfactant with creamy foam, milder than sulfates.
Effectiveness: Balances cleansing and mildness for daily use.
Pair With: Cocamidopropyl Betaine, Glycerin, Disodium Laureth Sulfosuccinate.
Avoid Using With: Hard water or high cationic surfactants.

Amphoteric Surfactants

These versatile surfactants have switchable behavior that makes them unusually compatible and gentle, boosting foam and reducing irritation.

Cocamidopropyl Betaine (1–5% face, up to 10% hair)

Best For: All skin types, sensitive skin.
What It Does: A coconut-derived surfactant that gently cleanses with soft lather, balancing pH.
Effectiveness: Less irritating than sulfates, great for daily use.
Pair With: Decyl Glucoside, Glycerin, Panthenol.
Avoid Using With: Harsh surfactants or very acidic ingredients.
Notes: Rare allergenicity from impurities is minimized in high-purity versions.

Cocamidopropyl Hydroxysultaine (1–3%)

Best For: Conditioning cleansers, all skin types.
What It Does: Gently cleanses, boosts foam, and conditions skin for a soft feel.
Effectiveness: Enhances primary surfactants, mild for daily use.
Pair With: Sodium Cocoyl Isethionate, Glycerin, Aloe Vera.
Avoid Using With: Very acidic formulas or strong anionic surfactants.

Other Mild Surfactants

Cocamide MIPA (Amide, 1–4%)

Best For: Foam quality, conditioning cleansers.
What It Does: A coconut-derived foam stabilizer and thickener, adds creamy texture.
Effectiveness: Enhances cleanser performance, used daily.
Pair With: SLES, Cocamidopropyl Betaine, Glycerin.
Avoid Using With: Very low pH or high acid/base concentrations.

PEG-7 Glyceryl Cocoate (Non-Ionic, 1–5%)

Best For: Dry, sensitive skin, micellar waters.
What It Does: Gently cleanses and conditions, leaving skin silky.
Effectiveness: Secondary surfactant for daily use, adds moisture.
Pair With: Disodium Laureth Sulfosuccinate, Glycerin.
Avoid Using With: Extreme pH (below 4 or above 9).
Notes: Contains PEG, which may form 1,4-Dioxane as an unlisted by-product, a potential concern for sensitive skin.

Stronger Anionic Surfactants
These are powerful cleansers that are best for oily skin but can be harsh for sensitive or dry skin.

Sodium Laureth Sulfate (SLES, concentration varies)

Best For: Normal to oily skin, deep cleansing.

What It Does: Creates abundant foam, removes oil and makeup effectively.
Effectiveness: Strong for daily use, less irritating than SLS.
Pair With: Glycerin, Betaine (moisturizing).
Avoid Using With: Sensitive skin actives (Retinol) right after cleansing.
Notes: Ethoxylated, may contain trace 1,4-Dioxane. Gentler than SLS but use with humectants for balance.

Sodium Lauryl Sulfate (SLS) (10–20%)

Best For: Oily skin, deep cleansing.
What It Does: Powerfully cleanses with rich foam, great for excess oil.
Effectiveness: Highly effective but can irritate at high doses.
Pair With: Hyaluronic Acid, Squalane (moisture balance).
Avoid Using With: Barrier-compromised or sensitive skin; pre-Retinoid use.
Notes: Irritation is dose-dependent; buffered formulas are safer.

Choosing the Right Surfactant for Your Skin

- **Sensitive/Dry Skin**: Pick mild options like Decyl Glucoside, Disodium Cocoyl Glutamate, or Sodium Cocoyl Glycinate to avoid irritation.
- **Oily Skin**: SLES or SLS (in lower doses with humectants) work for deep cleansing, but test for tolerance.
- **Combination Skin**: Try Cocamidopropyl Betaine or Sodium Methyl Cocoyl Taurate for balance.
- **Eco-Conscious**: Choose biodegradable surfactants like Coco Glucoside or Lauryl Glucoside.
- **Read Labels**: Look for gentle surfactants high on the ingredient list and avoid SLS/SLES if you have sensitive skin.
- **Patch-Test**: Test new cleansers on a small area, especially if using potent actives (retinol, vitamin C) afterward.

By choosing surfactants that match your skin type and pairing them wisely, you can cleanse effectively while keeping your skin healthy and comfortable.

Toners for Post-Cleanse Power

Toners refresh and soothe the skin, help remove residual impurities and excess oil, restore the skin's pH balance, and prepare it for better absorption of subsequent skincare products. Modern toners have evolved far beyond the harsh, alcohol-based astringents of the past. They now include hydrating essences, exfoliating acid treatments, and barrier-supporting formulations that address specific skin concerns while serving as an essential bridge between cleansing and moisturizing.

Toners are applied after cleansing to restore the skin's pH (typically 4.5–5.5), add lightweight hydration, deliver active ingredients, and prep the skin for serums or moisturizers. Today's toners are gentle, often alcohol-free, and range in texture from watery liquids to thicker essences. The following are the toner ingredients typically found in skincare products.

Denatured Alcohol (Alcohol Denat) (Solvent/Astringent, ≤5% routine, 5–20% targeted)

Best For: Oily skin, oil control.
What It Does: Reduces surface oil, tightens pores temporarily, and enhances ingredient penetration in classic toners.
Effectiveness: Quick drying; low doses for daily use, higher for targeted astringency.
Pair With: Glycerin, Hyaluronic Acid, Ceramides (offset dryness).
Avoid Using With: Sensitive/dry skin, high-concentration acids (Salicylic Acid), Retinoids.
Notes: Can increase dryness or sensitivity; opt for alcohol-free toners if prone to irritation.

Cucumis Sativus (Cucumber) Fruit Water (Hydrosol/Soothing, Base/Active)

Best For: Sensitive, sun-exposed skin.
What It Does: Cools, soothes irritation, reduces puffiness, and lightly hydrates.
Effectiveness: Immediate calming for daily AM/PM use.
Pair With: Aloe Vera, Panthenol (soothing), Glycerin (hydration).
Avoid Using With: Strong acids/Retinoids in same step (may sting).

Galactomyces Ferment Filtrate (Pitera™) (Fermentation/Brightening, 5–30%)

Best For: Dull skin, uneven tone, texture concerns.
What It Does: Brightens, hydrates, refines texture, and regulates sebum (K-beauty staple).
Effectiveness: Visible tone/texture improvement in 4–8 weeks.
Pair With: Niacinamide (brightening), Hyaluronic Acid, Lactic Acid (gentle exfoliation).
Avoid Using With: Very low pH actives in same step (less comfortable).
Notes: Hydration and brightness benefits are well-supported; sebum control varies.

Hamamelis Virginiana (Witch Hazel) Water (Alcohol-Free) (Astringent/Anti-inflammatory, Base/Active)

Best For: Oily skin, pore refining.
What It Does: Reduces oil, tightens pores temporarily, and soothes with anti-inflammatory benefits.
Effectiveness: Temporary astringency for daily AM/PM use.
Pair With: Chamomile, Aloe Vera, Glycerin (hydration).
Avoid Using With: Alcohol-heavy products, strong acids/Retinoids in same step.
Notes: Choose alcohol-free distillate to avoid dryness.

Rosa Damascena (Rose) Flower Water (Hydrosol/Soothing, Base/Active)

Best For: Sensitive skin, pH restoration.
What It Does: Lightly hydrates, soothes, and maintains skin's acidic pH with mild antioxidant benefits.

Effectiveness: Gentle hydration and calming for daily AM/PM use.
Pair With: Glycerin, Panthenol, Green Tea Extract (antioxidants).
Avoid Using With: Synthetic fragrances, harsh actives (overwhelm gentle benefits).
Notes: May contain natural fragrance allergens; use distilled grades for sensitive skin.

Tips for Choosing Toner Ingredients

- **Sensitive/Dry Skin**: Pick Cucumber or Rose Water for soothing hydration; avoid Alcohol Denat.
- **Oily Skin**: Use Witch Hazel (alcohol-free) or Galactomyces for oil control and refining.
- **Dull/Uneven Skin**: Choose Galactomyces with Niacinamide for brightening.
- **Check Labels**: Look for active ingredients high on the list and avoid irritants (Alcohol Denat) if sensitive.
- **Patch-Test**: Test toners, especially with actives (Retinoids), to ensure compatibility.

By selecting toners with ingredients suited to your skin, you'll prep your skin effectively for the rest of your routine, boosting results and keeping skin healthy.

HYDRATION HEROS

Healthy, hydrated skin depends on two key things: attracting water into the skin and keeping it there. This section explores the ingredients that do just that.

> **Humectants** act as moisture magnets, drawing water from the environment and deeper skin layers to plump and hydrate the surface.
>
> **Emollients and occlusives** work to smooth skin texture while creating a protective seal that prevents moisture loss and shields against environmental stressors.

Together, these ingredients are important for maintaining a strong, flexible skin barrier while preventing dryness, flaking, and irritation. Whether your skin is dry, dehydrated, or simply needs extra support during seasonal changes, these are some skincare product hydration heroes.

Humectants: Moisture Magnets

Your skin's ability to attract and retain moisture depends largely on humectants. These are specialized molecules that act like water magnets, drawing moisture from both the environment and the deeper layers of the skin to the surface. where it's needed most.

The best humectants closely resemble your skin's own natural moisturizing factor (NMF), a mix of water-loving compounds that help healthy skin hold onto moisture. When your NMF is balanced, your skin can maintain hydration even in dry, harsh, or otherwise challenging environments. However, factors such as aging, sun exposure, cold weather, or harsh skincare products can wear down this natural moisture network, leaving skin dry, rough, and sensitive.

It's important to note that topical humectants can backfire in low humidity conditions such as deserts and arid climates. When there isn't enough moisture in the air to attract, humectants may pull water upward from the deeper layers of your skin instead. This increases water loss from the skin's surface and can lead to dryness and irritation. That's why it's especially important in dry climates to pair humectants with occlusive ingredients, such as oils or butters, that lock moisture in and prevent evaporation.

Glycerin (Humectant, 2–10%)

Best For: All skin types, instant hydration.
What It Does: Pulls water to soften and hydrate skin, supporting barrier health.
Effectiveness: Immediate hydration, tacky at high levels; use daily AM/PM.
Pair With: Hyaluronic Acid, Ceramides, Niacinamide (barrier support).
Avoid Using With: Dry climates without occlusives, harsh alcohols.
Notes: Gold-standard humectant; avoid undiluted use to prevent stickiness.

Hyaluronic Acid (Humectant, 0.05–0.3%)

Best For: Plumping, anti-aging, deep hydration.
What It Does: Binds water for a plump, dewy look; reduces fine lines temporarily.
Effectiveness: Instant plumping, best sealed with moisturizer.
Pair With: Ceramides, Vitamin C, Peptides (anti-aging).
Avoid Using With: Dry environments without occlusives, strong acids (Salicylic Acid).
Notes: Layer with water-based products; excess can pull moisture from deeper layers.

Sodium Hyaluronate (Humectant, 0.05–0.3%)

Best For: Deep hydration, professional formulas.
What It Does: Salt form of hyaluronic acid; hydrates surface layers with better stability.
Effectiveness: Immediate hydration, best with emollients for lasting effect.
Pair With: Hyaluronic Acid, Peptides, Antioxidants.
Avoid Using With: Astringent alcohols (Witch Hazel).
Notes: Like hyaluronic acid, needs occlusives in dry climates to prevent TEWL.

Sodium PCA (NMF Component, 1–5%)

Best For: Dry/mature skin, natural hydration.
What It Does: Mimics skin's NMF to bind moisture, supporting barrier balance.
Effectiveness: Long-lasting hydration (24+ hours) when sealed with occlusives.
Pair With: Glycerin, Ceramides, Lactic Acid (NMF support).
Avoid Using With: Harsh surfactants, oil-only formulas.
Notes: Rare irritation risk for amino acid allergies.

Emollients and Occlusives

Emollients and occlusives are your skin's comforters and protectors. They don't just hydrate; they seal, soften, and strengthen. These ingredients provide lipid-based layer that smooths skin texture, locks in water, and defends against environmental stressors.

- **Emollients** work by filling in the microscopic cracks between skin cells, giving skin a smooth, supple feel. They provide nourishment through fatty acids, lipids, and antioxidants that help maintain your skin's elasticity and resilience. Plant oils, butters, beef tallow, silicones, and fatty alcohols are all examples of emollients.

- **Occlusives** work as shields. They form a physical barrier on the skin's surface to reduce transepidermal water loss (TEWL). Some, like shea butter and beeswax, still allow for a degree of healthy moisture exchange. Others, such as mineral oil and petrolatum, provide nearly total occlusion, which can be problematic.

Not all emollients and occlusives are created equal. Some are lightweight and fast-absorbing, making them ideal for oily or acne-prone skin. Others are rich and heavy to support dry, aging, or compromised skin that needs intensive barrier support. And there are a few that carry health concerns or allergy risks.

There are hundreds of emollient and occlusive ingredients used in skincare. These range from familiar plant oils like jojoba and almond oil, to luxury botanicals like marula and rosehip seed, to classic but controversial occlusives like mineral oil and petrolatum. The emollients and occlusives listed below are ones you're likely to find in products that support your skin's protective barrier.

Butyrospermum Parkii (Shea) Butter (Emollient/Occlusive, 5–20%)

Best For: Dry/mature skin, barrier support.
What It Does: Rich butter with fatty acids; nourishes and forms a breathable barrier.
Effectiveness: Hydrates and protects; improves barrier in 4–6 weeks.
Pair With: Glycerin, Ceramides, Vitamin E (antioxidants).
Avoid Using With: Oily or acne-prone skin; tree nut allergies (patch test).
Notes: Breathable unlike petrolatum; unrefined has more nutrients.

Caprylic/Capric Triglyceride (Emollient, 1–50%)

Best For: All skin types, non-comedogenic moisture.
What It Does: Coconut-derived, lightweight; softens and lightly seals moisture.
Effectiveness: Immediate softening, non-greasy; daily AM/PM use.
Pair With: Hyaluronic Acid, Niacinamide, Antioxidants.
Avoid Using With: Coconut/palm oil allergies (patch test).
Notes: Stable, well-tolerated even by acne-prone skin.

Ceramide NP (Barrier Lipid, 0.1–1%)

Best For: Dry skin, eczema, barrier repair.

What It Does: Restores skin's lipid barrier, locking in moisture for healthy skin.
Effectiveness: Improves hydration and barrier in 2–8 weeks with lipid blends.
Pair With: Cholesterol, Fatty Acids, Hyaluronic Acid.
Avoid Using With: None; optimized with lipid ratios (~3:1:1).
Notes: Best in multi-ceramide formulas for comprehensive barrier repair.

Dimethicone (Emollient/Occlusive, 0.5–20%)

Best For: All skin types, lightweight sealing.
What It Does: Silicone-based; forms breathable barrier, smooths texture.
Effectiveness: Instant smoothing and moisture lock; daily use.
Pair With: Hyaluronic Acid, Panthenol, Antioxidants.
Avoid Using With: Silicone sensitivities, fungal acne.
Notes: Non-comedogenic when formulated properly.

Petrolatum (Occlusive, concentration varies)

Best For: Severe dryness, post-procedure care.
What It Does: Forms near-impenetrable barrier, preventing 99% water loss.
Effectiveness: Immediate relief for dry skin; use sparingly under guidance.
Pair With: Humectants (apply first).
Avoid Using With: Acne-prone skin, actives needing penetration (Retinoids).
Notes: Avoid regular use; may disrupt natural moisture exchange.

Simmondsia Chinensis (Jojoba) Seed Oil (Emollient/Light Occlusive, 1–100%)

Best For: All skin types, sebum regulation.
What It Does: Mimics skin's sebum; lightweight, balances oil, and seals moisture.
Effectiveness: Hydrates without clogging; improves barrier in 4–6 weeks.
Pair With: Vitamin E, Rosehip Oil, Zinc (acne-prone skin).
Avoid Using With: Apply after water-based actives.
Notes: Acne-friendly, vegan-friendly in plant-derived forms.

Squalane (Emollient/Light Occlusive, 1–100%)

Best For: All skin types, natural moisture.
What It Does: Olive-derived; mimics skin lipids, softens, and lightly seals.
Effectiveness: Non-comedogenic, breathable; improves hydration in 4–6 weeks.
Pair With: Hyaluronic Acid, Ceramides, Retinol (reduces irritation).
Avoid Using With: None; apply after water-based actives.
Notes: Plant-derived versions ideal for vegan formulas.

Tips for Choosing Hydration Ingredients

- **Dry/Sensitive Skin:** Use hyaluronic acid, sodium PCA, shea butter, ceramide NP for gentle hydration and repair.
- **Oily/Acne-Prone Skin:** Choose squalane, jojoba oil, caprylic/capric triglyceride for lightweight, non-comedogenic moisture.

- **Mature Skin**: Pair hyaluronic acid with ceramides and squalane for anti-aging hydration.
- **Dry Climates**: Always seal humectants (glycerin, hyaluronic acid) with occlusives (shea butter, squalane).
- **Read Labels**: Look for humectants high on the list; avoid Petrolatum for acne-prone skin.
- **Patch-Test**: Test products with new ingredients, even moisturizers.

UV DEFENSE

Sunlight sustains life, but its ultraviolet (UV) rays are among the most damaging forces your skin will ever face. Prolonged, unprotected exposure to UVA rays penetrates deep into the skin where they break down collagen and elastin, while UVB rays cause surface damage, accelerating visible aging, causing hyperpigmentation, and increasing the risk of skin cancer. That's why UV defense ingredients are essential in skincare.

Many skincare products, such as moisturizers, primers, and foundations, include UV-filtering ingredients. The product may or may not have an SPF rating on the label. Products without an SPF rating will deliver some UV protection, but unlike regulated, SPF-rated products, the exact level of protection is unknown.

In the United States, adding SPF to skincare requires the product to comply with FDA regulations for OTC drugs, including formulation, testing, and labeling. Sun Protection Factor (SPF) measures how well a product protects against damaging UVB rays. The SPF rating reflects the time it takes for your skin to burn with sunscreen compared to without. For example, SPF 15 means it would take 15 times longer for your skin to burn with the sunscreen than without it.

UV-protective agents in skincare products work in two distinct ways:

- **Mineral** (physical) filters form a shield on the skin's surface, reflecting and scattering UV radiation. Common examples include zinc oxide and titanium dioxide, which provide broad-spectrum protection against both UVA and UVB rays.

- **Chemical** (organic) filters absorb UV energy and convert it into harmless heat. Chemical sunscreens can cause allergic reactions or irritation, especially in those with sensitive skin. Mineral-based sunscreens are often a better choice for those prone to redness, rosacea, or eczema.

Chemical sunscreen ingredients raise concerns. Some, including oxybenzone, avobenzone, octinoxate, and octocrylene, have been shown to enter the bloodstream at levels above the FDA's safety threshold under conditions of maximal use. Some have demonstrated endocrine-disrupting effects in animal studies.

Environmental impact is another important consideration. Oxybenzone and octinoxate have been linked to coral reef damage, leading to bans in some regions. Sunscreens labeled "reef-safe" are typically formulated with mineral filters like zinc oxide and titanium dioxide to minimize harm to marine ecosystems. The principle that "if it won't harm marine ecosystems, then it won't harm me" is worth considering when selecting sunscreen ingredients.

Avobenzone (Chemical UVA Filter, ≤3%)

Best For: UVA protection, anti-aging.
What It Does: Absorbs UVA rays to prevent deep skin damage and aging.
Effectiveness: Effective but photounstable; needs stabilizers (Octocrylene). Reapply every 2 hours.
Pair With: Octocrylene, Tinosorb S, Vitamin E (antioxidant).
Avoid Using With: Uncoated Titanium Dioxide/Zinc Oxide (degrades avobenzone).
Notes: Potential hormone disruption; systemic absorption concerns.

Octinoxate (Ethylhexyl Methoxycinnamate) (Chemical UVB Filter, ≤7.5%)

Best For: UVB protection, lightweight formulas.
What It Does: Blocks UVB rays to prevent sunburn; lightweight feel.
Effectiveness: Effective but photounstable; reapply every 2 hours.
Pair With: Avobenzone (stabilized), Vitamin E (antioxidant).
Avoid Using With: Unstabilized Avobenzone, retinoids (daytime).
Notes: Coral reef toxicity concerns (banned in Hawaii); possible hormone disruption.

Octisalate (Ethylhexyl Salicylate) (Chemical UVB Filter, ≤5%)

Best For: UVB protection, water-resistant formulas.
What It Does: Absorbs UVB rays, enhances water resistance, stabilizes other filters.
Effectiveness: Moderate UVB protection; stable in sunlight.
Pair With: Avobenzone, Octocrylene, Antioxidants.
Avoid Using With: Aspirin allergies (salicylate structure).
Notes: Mild irritation risk for sensitive skin.

Octocrylene (Chemical UVB/UVA-II Filter, ≤10%)

Best For: UVB protection, stabilizing avobenzone.
What It Does: Absorbs UVB/UVA-II, stabilizes other filters, water-resistant.
Effectiveness: Moderate protection boosts formula stability; reapply every 2 hours.
Pair With: Avobenzone, Antioxidants, water-resistant bases.
Avoid Using With: Sensitive skin (irritation risk), coral-safe formulas.
Notes: Potential endocrine concerns; environmental impact issues.

Oxybenzone (Benzophenone-3) (Chemical UVB/UVA Filter, ≤6%)

Best For: Broad-spectrum protection, SPF boosting.
What It Does: Absorbs UVB/UVA, stabilizes other filters.
Effectiveness: Good broad-spectrum coverage; reapply every 2 hours.

Pair With: Avobenzone, Octisalate, Antioxidants.
Avoid Using With: Sensitive skin, coral reefs (banned in some regions).
Notes: Higher allergy risk, hormone disruption concerns.

Titanium Dioxide (Mineral UV Filter, 5–25%)

Best For: Sensitive skin, broad-spectrum protection.
What It Does: Reflects/scatters UVB/UVA-II, minimal UVA-I; forms a physical shield.
Effectiveness: Immediate protection, stability; less UVA-I coverage than zinc.
Pair With: Zinc Oxide, Iron Oxides (tinted), Glycerin (hydration).
Avoid Using With: Unstabilized Avobenzone (degradation risk).
Notes: May leave white cast; nano forms improve aesthetics.

Zinc Oxide (Mineral UV Filter, 2–25%)

Best For: Sensitive/acne-prone skin, broad-spectrum protection.
What It Does: Reflects/scatters UVA/UVB, soothes inflammation, antimicrobial.
Effectiveness: Immediate, broadest protection; reduces redness in days.
Pair With: Titanium Dioxide, Niacinamide, Panthenol (soothing).
Avoid Using With: Strong acids (Ascorbic Acid) that alter properties.
Notes: Gold standard for sensitive skin; reef-safe, minimal white cast in modern formulas.

Tips for Choosing UV Defense Ingredients

- **Sensitive Skin**: Use Zinc Oxide or Titanium Dioxide for gentle, broad-spectrum protection.
- **Oily/Acne-Prone Skin**: Choose Zinc Oxide (non-comedogenic, antimicrobial) or lightweight chemical filters like Octisalate.
- **Anti-Aging**: Prioritize Avobenzone or Zinc Oxide for strong UVA protection.
- **Reef-Safe**: Opt for Zinc Oxide/Titanium Dioxide to avoid environmental harm.
- **Read Labels**: Ensure broad-spectrum coverage; look for stabilizers (Octocrylene) with chemical filters.
- **Reapply**: Every 2 hours for chemical filters, less critical for minerals.

TARGETED TREATMENT ACTIVES

Once your skin's foundational needs are met, cleansing, hydration, and barrier support, targeted treatment actives take your routine to the next level. These high-performance ingredients are designed to correct specific concerns such as uneven tone, breakouts, inflammation, fine lines, and loss of firmness. This section covers:

- **Exfoliants** with scrubs, enzymes, and chemicals like AHAs, BHAs, and PHAs help renew the skin by increasing cell turnover and clearing congestion.

- **Antioxidants** that defend against free radical damage and form the backbone of anti-aging strategies.

- **Brightening agents** that address uneven skin tone, dark spots, and dullness through various mechanisms like melanin inhibition and gentle exfoliation.

- **Anti-acne ingredients** that tackle breakouts by regulating oil production, unclogging pores, and reducing bacterial overgrowth.

- **Soothing and anti-inflammatory agents** that calm irritated skin and support healing.

Used strategically, these ingredients can deliver transformative results. This section will help you understand how they work, what to combine, or avoid, and how to choose the right targeted treatment actives for your unique skin needs.

Exfoliants

The outermost layer of your skin, the stratum corneum, naturally sheds about every 28 days in your 20s. This process slows with age, sun damage, and certain skin conditions, resulting in a buildup of dead cells that can cause dullness, flakiness, clogged pores, and uneven texture.

Exfoliants can act as a reset button for your skin. By removing dead cells from the surface, they help unclog pores, smooth texture, brighten tone, and allow active ingredients to penetrate more effectively. But not all exfoliants work the same way or suit every skin type. Exfoliants function in three main ways:

- **Physical exfoliation** uses particles or tools to manually remove dead skin cells through scrubbing or friction. Common ingredients include sugar, salt, jojoba beads, and crushed walnut shells.

- **Mechanical exfoliation** uses tools such as brushes, washcloths, or exfoliating gloves to physically buff away dead skin. These work best after soaking or steaming the skin to loosen dirt, oil, and debris trapped in pores.

- **Chemical exfoliation** uses acids and enzymes to dissolve the bonds holding dead cells together so they naturally shed. Common acid exfoliants include alpha hydroxy acids (AHAs), like glycolic and lactic acid, which work on the skin's surface; beta hydroxy acids (BHAs), like salicylic acid, which penetrate pores; and polyhydroxy acids (PHAs), which are gentler and suitable for sensitive skin. Enzymatic exfoliants include papain (from papaya), bromelain (from pineapple), and bacterial enzymes like subtilisin, which provide a milder alternative to acids and are particularly suited to sensitive skin types.

Which exfoliation method is best for your skin depends on your skin type, condition, sensitivity, and specific concerns. The right frequency and the right exfoliant can dramatically transform skin health by improving texture, enhancing product absorption, regulating oil production, minimizing the appearance of fine lines, and creating a more radiant complexion.

The wrong exfoliation method - or overdoing it -can lead to irritation, inflammation, and barrier damage. Whether you're scanning a label or building a routine, this section will help you make more informed choices with clarity and confidence.

Avena Sativa (Oat) Kernel Flour (Physical Exfoliant, 5–15%)

Best For: Sensitive skin, soothing exfoliation.
What It Does: Gently buffs dead cells; oat β-glucans calm irritation.
Effectiveness: Mild smoothing 1–2×/week; suits reactive skin.
Pair With: Honey, Hyaluronic Acid, Calendula (soothing).
Avoid Using With: Strong chemical exfoliants, hot water (increases irritation).
Notes: Anti-inflammatory; ideal for redness-prone skin.

Bromelain (Enzymatic Exfoliant, Varies)

Best For: Sensitive/redness-prone skin, gentle renewal.
What It Does: Pineapple enzyme dissolves dead cell bonds, calms skin.
Effectiveness: Smoothness after 1–2 weekly uses; minimal irritation.
Pair With: Aloe Vera, Panthenol, Glycerin (hydration).
Avoid Using With: High heat, extreme pH, strong acids/scrubs (same session).
Notes: Gentle, natural enzyme for low-irritation exfoliation.

Citric Acid (Alpha Hydroxy Acid, 3–8%)

Best For: Gentle exfoliation, brightening, sensitive skin.
What It Does: Citrus-derived AHA smooths and brightens with antioxidant benefits.
Effectiveness: Texture/tone improves in 3–5 weeks at pH 3.5–4.5.
Pair With: Hyaluronic Acid, Vitamin E, Allantoin (soothing).
Avoid Using With: High-level acids, retinoids, L-Ascorbic Acid (pH mismatch).
Notes: Less irritating than glycolic acid; good entry-level AHA.

Glycolic Acid (Alpha Hydroxy Acid, 5–15%)

Best For: Rapid texture improvement, anti-aging, brightening.
What It Does: Smallest AHA; deeply exfoliates, boosts cell turnover.
Effectiveness: Visible results in 1–3 weeks at pH <4; potent.
Pair With: Hyaluronic Acid, Niacinamide, Antioxidants.
Avoid Using With: Retinoids, Benzoyl Peroxide, strong acids (same routine).
Notes: Can irritate sensitive skin; start low and build tolerance.

Lactic Acid (Alpha Hydroxy Acid, 5–12%)

Best For: Dry/sensitive skin, hydration + exfoliation.
What It Does: Gently exfoliates, boosts moisture via NMF support.
Effectiveness: Smoother, hydrated skin in 2–4 weeks at pH 3.5–4.5.

Pair With: Niacinamide, Ceramides, Hyaluronic Acid.
Avoid Using With: Strong retinoids, high-level acids (same application).
Notes: Gentler than glycolic; ideal for dry or sensitive skin.

Mandelic Acid (Alpha Hydroxy Acid, 5–10%)

Best For: Sensitive skin, acne, hyperpigmentation, deeper tones.
What It Does: A large molecule AHA that gently exfoliates, fights bacteria.
Effectiveness: Clears pores, evens tone in 4–8 weeks; low irritation.
Pair With: Niacinamide, Hyaluronic Acid, Bisabolol (soothing).
Avoid Using With: High-level acids, L-Ascorbic Acid (pH issues).
Notes: Antibacterial; great for acne and darker skin tones.

Papain (Enzymatic Exfoliant, Varies)

Best For: Sensitive skin, gentle renewal.
What It Does: Papaya enzyme dissolves dead cell bonds, smooths skin.
Effectiveness: Visible smoothness after 1–2 weekly uses.
Pair With: Aloe Vera, Glycerin, Allantoin (soothing).
Avoid Using With: High heat, extreme pH, strong acids/scrubs (same session).
Notes: Gentle enzyme; ideal for irritation-prone skin.

Salicylic Acid (Beta Hydroxy Acid, 0.5–2%)

Best For: Acne-prone/oily skin, pore clearing.
What It Does: Oil-soluble BHA clears pores, reduces inflammation.
Effectiveness: Acne/comedone reduction in 1–2 weeks; potent.
Pair With: Niacinamide, Hyaluronic Acid, Green Tea Extract.
Avoid Using With: Benzoyl Peroxide, high-level acids, L-Ascorbic Acid (same routine).
Notes: Gold standard for acne; may irritate sensitive skin.

Sucrose (Natural Physical Exfoliant)

Best For: Gentle exfoliation, natural skincare, sensitive skin.
What It Does: Sugar's fine, rounded granules gently buff away dead skin; trace glycolic acid adds mild chemical exfoliation. Dissolves during use for a soft finish.
Effectiveness: Immediate smoothing at 10–30% in scrubs; kind to most skin types 1–2×/week.
Pair With: Honey, Glycerin (hydration), Jojoba Oil, Chamomile (soothing).
Avoid Using With: Strong acids (Glycolic Acid), active acne, hot water (dissolves early).
Notes: Biodegradable, gentler than nut shells; ideal for eco-friendly formulas.

Urea (Keratolytic/Humectant, 2–20%)

Best For: Dry/rough skin, keratosis pilaris.
What It Does: Softens rough texture, hydrates at low levels, exfoliates at higher levels.
Effectiveness: Smoother skin in 2–4 weeks; 10%+ for keratolysis.
Pair With: Glycerin, Lactic Acid, Ceramides (barrier support).
Avoid Using With: High-level exfoliants, strong preservatives (irritation risk).

Notes: Enhances active penetration; monitor high concentrations.

Tips for Choosing the Right Exfoliant

- **Sensitive Skin:** Use Lactic Acid, Mandelic Acid, Bromelain, or Papain for gentle renewal.
- **Oily/Acne-Prone Skin:** Choose Salicylic Acid or Mandelic Acid for pore-clearing.
- **Dry Skin:** Opt for Lactic Acid or Urea for hydration + exfoliation.
- **Frequency:** Start 1–2×/week; chemical/enzymatic can increase to 3–4×/week if tolerated.
- **Check Labels:** Look for pH (3.5–4.5 for acids), concentration, and soothing ingredients.
- **Patch-Test:** Test new exfoliants to avoid irritation, especially with acids.

Antioxidants

Think of your skin as a fortress under constant siege. Environmental aggressors like pollution, UV radiation, and even your body's own metabolic processes launch daily attacks in the form of free radicals.

Free radicals are unstable molecules, hungry to steal electrons from your skin's vital structures. This electron theft, known as oxidation, gradually damages collagen, elastin, and even DNA. This leads to visible signs of premature aging, including wrinkles, dark spots, and sagging skin.

Antioxidants serve as your skin's defensive army against this oxidative damage. They selflessly donate electrons to neutralize free radicals without becoming unstable themselves, preserving your skin's structural integrity and youthful appearance.

In skincare, antioxidants do more than just protect your skin; they also support skin repair, brighten tone, calm irritation, and enhance the performance of other active ingredients. Each antioxidant has a unique mechanism and benefit profile.

Whether you're targeting dullness, sensitivity, or premature aging, antioxidants are essential allies in building resilient, radiant skin. Understanding how to effectively include these protective ingredients in your skincare routine can make a significant difference in maintaining skin health and preventing premature aging.

Ascorbic Acid (L-Ascorbic Acid) (Vitamin C, 5–20%)

Best For: Brightening, anti-aging, collagen support.
What It Does: Neutralizes free radicals, brightens tone, boosts collagen.
Effectiveness: Visible brightening in 2–4 weeks, firmness in 8–12 weeks at pH ≤3.5.
Pair With: Vitamin E, Ferulic Acid (stability), Sunscreen (daytime).
Avoid Using With: Benzoyl Peroxide (oxidizes), high acids (same step).
Notes: Light/oxygen-sensitive; use airless packaging.

3-O-Ethyl Ascorbic Acid (Vitamin C Derivative, 1–3%)

Best For: Sensitive skin, stable vitamin C benefits.
What It Does: Brightens, protects against free radicals, gentler than L-ascorbic acid.
Effectiveness: Brightening in 4–8 weeks; stable at wider pH range.
Pair With: Vitamin E, Niacinamide, Hyaluronic Acid.
Avoid Using With: Benzoyl Peroxide (same step).
Notes: More stable, less irritating than pure vitamin C.

Arbutin (Brightening Antioxidant, 0.5–2%)

Best For: Sensitive skin, hyperpigmentation.
What It Does: Inhibits melanin, reduces dark spots, provides antioxidant support.
Effectiveness: Tone evening in 8–12 weeks; gentle.
Pair With: Vitamin C, Niacinamide, Hyaluronic Acid.
Avoid Using With: High acids/scrubs (sensitive skin).
Notes: Store away from heat/light to prevent breakdown.

Azelaic Acid (Multitasking Antioxidant, 8–20%)

Best For: Acne, rosacea, hyperpigmentation, sensitive skin.
What It Does: Exfoliates, brightens, reduces bacteria/inflammation.
Effectiveness: Improves acne/PIH in 4–12 weeks; pregnancy-safe.
Pair With: Niacinamide, Hyaluronic Acid, Gentle Retinoids (alternate).
Avoid Using With: High acids (same application).
Notes: Gentle, effective for redness and pigmentation.

Camellia Sinensis Leaf Extract (Green Tea) (Polyphenol Antioxidant, 0.5–3%)

Best For: Redness, oil control, environmental protection.
What It Does: Calms inflammation and fights free radicals, antimicrobial.
Effectiveness: Reduces redness/oxidative stress in 4–8 weeks.
Pair With: Vitamin C, Niacinamide, Hyaluronic Acid.
Avoid Using With: Overly complex routines (sensitive skin).
Notes: Needs light-protected packaging for EGCG stability.

Ferulic Acid (Antioxidant Enhancer, 0.5–1%)

Best For: Stabilizing antioxidants, environmental protection.
What It Does: Boosts vitamins C and E, neutralizes free radicals.
Effectiveness: Enhances photoprotection; benefits in weeks with C/E.
Pair With: Vitamin C, Vitamin E, Sunscreen.
Avoid Using With: High-pH products (reduces stability).
Notes: Key in "CE Ferulic" serums for synergy.

Niacinamide (Vitamin B3, 2–10%)

Best For: Strengthening skin barrier, controlling oil, brightening, reducing redness.

What It Does: Enhances ceramide production, calms inflammation, evens skin tone, minimizes pores.
Effectiveness: Visible improvement in texture and redness within 4–8 weeks.
Pair With: Ceramides, Hyaluronic Acid, Vitamin C (stable formulas).
Avoid Using With: High L-Ascorbic Acid (same step, if flushing occurs).
Notes: Highly versatile; compatible with vitamin C in modern, stable formulas, debunking outdated myths.

Retinol (Vitamin A, 0.25–1%)

Best For: Anti-aging, texture, tone improvement.
What It Does: Speeds turnover, boosts collagen, fights free radicals.
Effectiveness: Reduces lines/texture in 12–16 weeks.
Pair With: Hyaluronic Acid, Niacinamide, Sunscreen.
Avoid Using With: Benzoyl Peroxide, strong acids (same step).
Notes: Start 2–3 nights/week; increases photosensitivity.

Tocopherol (Vitamin E) (Antioxidant, 0.5–5%)

Best For: Barrier protection, soothing, antioxidant synergy.
What It Does: Protects lipids, reduces inflammation, boosts C/E.
Effectiveness: Enhances resilience with consistent use.
Pair With: Vitamin C, Ferulic Acid, Sunscreen.
Avoid Using With: High levels in oily skin (may feel heavy).
Notes: Mixed tocopherols broaden protection.

Tips for Choosing Antioxidants

- **Sensitive Skin**: Use arbutin, azelaic acid, 3-o-ethyl ascorbic acid for gentle benefits.
- **Acne-Prone Skin**: Choose niacinamide, azelaic acid for oil control and clarity.
- **Anti-Aging:** prioritize retinol, ascorbic acid, ferulic acid for collagen and firmness.
- **Brightening**: I=Opt for arbutin, niacinamide, vitamin C for even tone.
- **Read Labels**: Look for stable packaging (opaque/airless) and synergistic pairings.
- **Patch-Test**: Always patch test potent actives (retinol, ascorbic acid) to avoid irritation.

Brightening Agents

Brightening agents target one of the most common complexion concerns: uneven tone. Discoloration can be caused by excessive UV and visible light exposure, post-inflammatory changes after acne or irritation, hormonal shifts (including melasma), heat, friction, and even pollution. The result is skin that looks dull, blotchy, or older than it is.

Brightening skin isn't about bleaching it; it's about restoring balance and clarity so that your skin's natural luminosity shows through. These ingredients work via complementary pathways: some dial down melanin production by inhibiting key enzymes like tyrosinase (and related TRP enzymes); others reduce melanosome transfer from melanocytes to keratinocytes. They often act as gentle exfoliants to lift away pigmented surface cells, serve as anti-inflammatories to calm pigmentation triggers, and as antioxidants to intercept the oxidative stress that sparks new discoloration.

Choosing the right brightening agent strategy means matching agent mechanisms to the concern and your skin's temperament. Sensitive or deeper skin tones often do best with non-irritating inhibitors and anti-inflammatories to avoid rebound hyperpigmentation, while more resilient skin may tolerate short, guided use of stronger actives.

Consistency matters: most see meaningful improvement over 6 to 12 weeks with a layered routine that includes a broad-spectrum SPF every morning, a melanin inhibitor, an antioxidant, and a gentle exfoliant used thoughtfully. The goal is an even, luminous complexion that's maintained, not a quick, harsh fix that upsets the barrier or stokes more pigment.

3-O-Ethyl Ascorbic Acid (Vitamin C Derivative, 1–3%)

Best For: Sensitive skin, stable brightening.
What It Does: Brightens via melanin inhibition, provides antioxidant protection, gentler than L-ascorbic acid.
Effectiveness: Reduces pigmentation in 4–8 weeks; stable in formulas.
Pair With: Vitamin E, Niacinamide, Kojic Acid (synergy).
Avoid Using With: Benzoyl Peroxide (reduces efficacy).
Notes: Oil-soluble, less irritating, ideal for sensitive skin.

Alpha-Arbutin (Tyrosinase Inhibitor, 1–2%)

Best For: Stubborn pigmentation, melasma, sensitive skin.
What It Does: Inhibits melanin production, fades dark spots gently.
Effectiveness: Visible lightening in 8–12 weeks; stable vs. hydroquinone.
Pair With: Vitamin C, Niacinamide, Lactic Acid (exfoliation).
Avoid Using With: High-pH products (>7, destabilizes).
Notes: Gentle, effective for long-term use.

Ascorbic Acid (L-Ascorbic Acid) (Vitamin C, 10–20%)

Best For: Comprehensive brightening, anti-aging.
What It Does: Inhibits melanin, disperses pigment, boosts collagen.
Effectiveness: Brightens in 2–4 weeks at pH ≤3.5; potent but unstable.
Pair With: Vitamin E, Ferulic Acid, Sunscreen.
Avoid Using With: Benzoyl Peroxide, high acids (same step).
Notes: Needs airless packaging; patch-test for sensitivity.

Azelaic Acid (Multitasking Brightener, 8–20%)

Best For: Acne, rosacea, hyperpigmentation, sensitive skin.
What It Does: Exfoliates, brightens, reduces inflammation/bacteria.
Effectiveness: Improves tone/acne in 4–12 weeks; pregnancy-safe.
Pair With: Niacinamide, Hyaluronic Acid, Gentle Retinoids (alternate).
Avoid Using With: High acids (same application).
Notes: Gentle, versatile for sensitive skin.

Glycyrrhiza Glabra (Licorice Root Extract) (Brightening Antioxidant, 0.5–5%)

Best For: Sensitive skin, melasma, redness reduction.
What It Does: Inhibits tyrosinase, calms inflammation, brightens tone.
Effectiveness: Reduces pigmentation/redness in 4–8 weeks.
Pair With: Niacinamide, Vitamin C, Arbutin.
Avoid Using With: Benzoyl Peroxide (may deactivate).
Notes: UV-sensitive; use opaque packaging or at night.

Kojic Acid (Fermentation-Derived Brightener, 1–4%)

Best For: Dark spots, melasma, acne-prone skin.
What It Does: Inhibits tyrosinase, lightens pigmentation, antimicrobial.
Effectiveness: Visible results in 8–12 weeks; may irritate sensitive skin.
Pair With: Glycolic Acid, Vitamin C, Glycerin (hydration).
Avoid Using With: Benzoyl Peroxide, high retinoids/acids (same step).
Notes: Patch test; especially for sensitive skin; use sunscreen to prevent photosensitivity.

Niacinamide (Vitamin B3, 2–10%)

Best For: Strengthening skin barrier, controlling oil, brightening, reducing redness.
What It Does: Enhances ceramide production, calms inflammation, helps even skin tone, minimizes pores.
Effectiveness: Visible improvement in texture and redness within 4–8 weeks.
Pair With: Ceramides, Hyaluronic Acid, Vitamin C (stable formulas).
Avoid Using With: High L-Ascorbic Acid (same step, if flushing occurs).
Notes: Highly versatile; compatible with vitamin C in modern, stable formulas, debunking outdated myths.

Retinol (Vitamin A, 0.25–1%)

Best For: Anti-aging, texture, tone improvement.
What It Does: Speeds turnover, boosts collagen, fights free radicals.
Effectiveness: Reduces lines/texture in 12–16 weeks.
Pair With: Hyaluronic Acid, Niacinamide, Sunscreen.
Avoid Using With: Benzoyl Peroxide, strong acids (same step).

Tetrahexyldecyl Ascorbate (Vitamin C Derivative, 1–5%)

Best For: Oil-based brightening, deep penetration.

What It Does: Brightens, protects with superior stability, penetrates well.
Effectiveness: Reduces pigmentation in 4–8 weeks; oil-soluble.
Pair With: Vitamin E, Niacinamide, lipid-based actives.
Avoid Using With: Water-based formulas (poor solubility).
Notes: Stable, effective in serums/creams.

Tranexamic Acid (Pigmentation Inhibitor, 2–5%)

Best For: Melasma, stubborn PIH, long-term maintenance.
What It Does: Blocks melanocyte-cell interaction reduces pigmentation.
Effectiveness: Improves melasma/PIH in 8–12 weeks; no rebound.
Pair With: Niacinamide, Vitamin C, Sunscreen.
Avoid Using With: High acids (same step, sensitive skin).
Notes: Ideal for resistant pigmentation; safe for long-term use.

Tips for Choosing Brightening Agents

- **Sensitive Skin**: Choose arbutin, azelaic acid, niacinamide for gentle action.
- **Stubborn Pigmentation**: Choose tranexamic acid, alpha-arbutin, kojic acid.
- **Anti-Aging**: Opt for ascorbic acid, tetrahexyldecyl ascorbate for collagen benefits.
- **Sunscreen Essential**: Pair all brighteners with SPF to prevent new pigmentation.
- **Check Labels**: Look for stable packaging, pH (for vitamin C), and concentrations.
- **Patch-Test**: Test potent actives (kojic acid, ascorbic acid) to avoid irritation.

Anti-Acne Agents

Acne isn't just a teenage rite of passage; it's a chronic, relapsing skin condition that can affect any age and skin type. Whether it's hormonal, stress-induced, or triggered by product buildup, breakouts usually stem from overlapping factors: excess oil, clogged pores, inflammation, and microbial imbalance (most notably Cutibacterium acnes and, in some cases, Malassezia yeast).

This section focuses on non-exfoliant, non-antioxidant, non-brightening actives that directly target acne's core pathways: think sebum regulators, antimicrobials, anti-inflammatories, and barrier-supportive agents. (You'll find exfoliants, antioxidants, and tone-evening brighteners in their own dedicated sections - pair them thoughtfully with the agents below rather than stacking everything at once.)

When addressing acne, choose one exfoliant (see the Exfoliants section) as the primary resurfacer; for example, salicylic acid for pore decongestion or azelaic acid for gentle keratolysis plus redness control. Be sure to introduce them slowly.

Reserve stronger AHAs for texture or post-blemish marks and keep PHAs/enzymes for very reactive skin. Avoid layering multiple strong acids with retinoids or benzoyl peroxide in the same application if you're prone to irritation.

Building an acne plan with these agents:

- **Reduce Sebum**: Niacinamide, zinc, green tea (EGCG), prescription topical clocortolone.
- **Target Microbes**: benzoyl peroxide (broad antibacterial), sulfur, resorcinol, topical prescription antibiotics, antifungals for Malassezia.
- **Calm Inflammation**: Licorice, centella, allantoin, colloidal oatmeal; short, strategic use of salicylic or azelaic acids (from the Exfoliants section) can also help.
- **Protect Your Skin Barrier**: Ceramides, cholesterol/fatty acids, panthenol. Always pair actives with a non-pore-clogging moisturizer and daily SPF to prevent post-blemish marks.

Keep changes incremental, track tolerance, and prioritize consistency over intensity. The right mix of anti-acne agents, plus smart use of exfoliants and brighteners (covered elsewhere), delivers clearer skin with fewer setbacks.

Be Mindful of Nuances
Benzoyl peroxide can bleach fabrics. Many actives increase photosensitivity. "Comedogenic ratings" depend on the full formula, not a single ingredient. If you're pregnant or highly sensitive, azelaic acid and sulfur are typically better-tolerated options, while retinoids should be avoided.

With the right mix and consistent, gentle use, you can reduce flare-ups, shorten healing time, minimize marks, and support a clearer, calmer, more resilient complexion without sacrificing the skin barrier.

A Note About Malassezia ("Fungal Acne")
Unlike classic bacterial acne, Malassezia (aka *Pityrosporum folliculitis*) is a yeast overgrowth in hair follicles. Clues include tiny, uniform, often itchy bumps (1–3 mm) that cluster on the forehead, hairline, chest, back, or shoulders that worsen with heat, sweat, occlusive products, or after antibiotics/steroids. Blackheads are uncommon.

Malassezia management centers on antifungals (like ketoconazole, selenium sulfide, or zinc pyrithione) and yeast-friendly routines. This means using lightweight formulas, simple emollients (like squalane), and avoiding heavy oils and esters. If breakouts are itchy, uniform, and resistant to standard acne care, consult a dermatologist for proper diagnosis.

Activated Charcoal (Absorbent, 1–5%)

Best For: Oily skin, pore detox.
What It Does: Absorbs oil/impurities, cleanses pores.
Effectiveness: Immediate oil reduction; use 1–2×/week.
Pair With: Hyaluronic Acid, Aloe Vera, Glycerin.
Avoid Using With: Daily use (dry/sensitive skin), chemical exfoliants (same session).
Notes: Rinse thoroughly; avoid overuse.

Azelaic Acid (Multitasking Acid, 8–20%)

Best For: Acne, rosacea, sensitive skin, PIH.
What It Does: Exfoliates, kills bacteria, reduces inflammation, brightens.
Effectiveness: Improves acne/PIH in 4–12 weeks; pregnancy-safe.
Pair With: Niacinamide, Hyaluronic Acid, Gentle Retinoids (alternate).
Avoid Using With: High acids (same application).
Notes: Gentle, versatile; ideal for sensitive skin.

Benzoyl Peroxide (Antibacterial, 2.5–10%)

Best For: Inflammatory acne, bacterial control.
What It Does: Kills P. acnes bacteria with oxygen, reduces inflammation, mildly exfoliates.
Effectiveness: Reduces acne in 2–4 weeks; 2.5% as effective as higher strengths with less irritation.
Pair With: Hyaluronic Acid, Ceramides, Niacinamide (soothing).
Avoid Using With: Retinoids, Vitamin C (same step, oxidizes).
Notes: Use sunscreen; start with lower concentrations.

Kaolin (Gentle Clay, 3–15%)

Best For: Sensitive/combination skin, mild cleansing.
What It Does: Absorbs oil, gently purifies pores.
Effectiveness: Mild cleansing; use 1–2×/week.
Pair With: Chamomile, Glycerin, Light Moisturizers.
Avoid Using With: Very dry skin (may dry out).
Notes: Gentler than bentonite; suits regular use.

Melaleuca Alternifolia (Tea Tree Oil) (Natural Antimicrobial, 5%)

Best For: Mild/moderate acne, natural treatments.
What It Does: Kills P. acnes, reduces redness, anti-inflammatory.
Effectiveness: Reduces acne in 4–12 weeks; slower but gentler.
Pair With: Niacinamide, Aloe Vera, Witch Hazel.
Avoid Using With: Retinoids, Benzoyl Peroxide (same step, irritation).
Notes: Dilute properly; patch-test to avoid sensitivity.

Niacinamide (Vitamin B3, 2–10%)

Best For: Strengthening skin barrier, controlling oil, brightening, reducing redness.
What It Does: Enhances ceramide production, calms inflammation, evens skin tone, minimizes pores.
Effectiveness: Visible improvement in texture and redness within 4–8 weeks.
Pair With: Ceramides, Hyaluronic Acid, Vitamin C (stable formulas).
Avoid Using With: High L-Ascorbic Acid (same step, if flushing occurs).
Notes: Highly versatile; compatible with vitamin C in modern, stable formulas, debunking outdated myths.

Retinol (Vitamin A, 0.25–1%)

Best For: Anti-aging, texture, tone improvement.
What It Does: Speeds turnover, boosts collagen, fights free radicals.
Effectiveness: Reduces lines/texture in 12–16 weeks.
Pair With: Hyaluronic Acid, Niacinamide, Sunscreen.
Avoid Using With: Benzoyl Peroxide, strong acids (same step).

Salicylic Acid (Beta Hydroxy Acid, 0.5–2%)

Best For: Oily/acne-prone skin, pore clearing.
What It Does: Exfoliates inside pores, reduces oil, calms inflammation.
Effectiveness: Clears comedones/acne in 1–2 weeks; potent at pH 3–4.
Pair With: Niacinamide, Hyaluronic Acid, Green Tea Extract.
Avoid Using With: Benzoyl Peroxide, high acids (same step).
Notes: Gold standard for acne; patch-test for sensitivity.

Sulfur (Natural Mineral, 3–10%)

Best For: Mild acne, sensitive/rosacea-prone skin.
What It Does: Kills bacteria, exfoliates, reduces oil/inflammation.
Effectiveness: Reduces acne in 1–3 weeks; gentler than benzoyl peroxide.
Pair With: Niacinamide, Glycerin, Zinc (oil control).
Avoid Using With: Metal-containing products (may stain).
Notes: Mild odor; ideal for sensitive skin.

Zinc PCA (Oil Control, 0.1–1%)

Best For: Oily/acne-prone skin, sebum regulation.
What It Does: Reduces oil, mild antibacterial action.
Effectiveness: Improves oil control in 2–4 weeks; supportive role.
Pair With: Niacinamide, Panthenol, Salicylic Acid.
Avoid Using With: Highly acidic products (disrupts zinc).
Notes: Gentle, enhances other actives.

Tips for Choosing Anti-Acne Ingredients

- **Sensitive Skin**: Use Azelaic Acid, Sulfur, Niacinamide for gentle action.
- **Oily Skin**: Choose Salicylic Acid, Benzoyl Peroxide, Zinc PCA.
- **Inflammatory Acne**: Prioritize Benzoyl Peroxide, Tea Tree Oil.
- **Non-Inflammatory Acne**: Opt for Salicylic Acid, Azelaic Acid.
- **Frequency**: Start 1–2×/week for potent actives; increase if tolerated.
- **Patch-Test**: Always patch test strong actives like benzoyl peroxide and salicylic acid to avoid irritation.

Soothing and Anti-Inflammatory Agents

When skin turns red, stings, or becomes inflamed, it's usually signaling barrier stress - often from environmental exposure (UV, pollution, wind), over-exfoliation, harsh products, temperature shifts, or microbiome imbalance. Calming and anti-inflammatory actives help interrupt that stress response: they reduce redness and discomfort, soothe visible irritation, and support the skin's natural repair processes.

These ingredients are especially valuable for sensitive, reactive, or post-procedure skin, where comfort and barrier integrity are non-negotiable. By restoring balance and resilience, they make routines more tolerable and create a stable foundation so other actives (like retinoids or acids) can work more effectively with fewer side effects.

Soothing and anti-inflammatory ingredients work through complementary mechanisms: reinforcing the skin barrier (ceramide/lipid support and reduced TEWL), modulating inflammatory signals, quenching oxidative stress, calming neurosensory irritation, balancing the microbiome, and easing visible flushing. Together, they're essential allies for anyone seeking skin that is calmer, stronger, and less reactive day to day.

Allantoin (Skin-Calming Diureide, 0.5–2%)

Best For: Sensitive skin, post-procedure recovery, minor wounds.
What It Does: Calms irritation, promotes gentle exfoliation, supports moisture barrier.
Effectiveness: Reduces irritation and aids healing within days; visible results in 1–2 weeks.
Pair With: Ceramides, Panthenol, Glycerin (barrier repair).
Avoid Using With: High Exfoliating Acids, Retinoids (same step, sensitive skin).
Notes: Gentle, widely tolerated; ideal for post-procedure care.

Aloe Barbadensis Leaf Juice (Aloe Vera) (Hydrating Botanical, 5–20%)

Best For: Sensitive/reactive skin, sunburn, minor irritations.
What It Does: Hydrates, calms inflammation, provides antioxidant benefits.
Effectiveness: Immediate soothing; long-term benefits in 2–4 weeks with ≥10% concentration.
Pair With: Allantoin, Hyaluronic Acid, Niacinamide.
Avoid Using With: Alcohol-based products, strong acids (unbuffered).
Notes: Check for high aloe content on labels; avoid low-potency extracts.

Avena Sativa (Oat) Kernel Extract (Anti-Inflammatory Oat Extract, 0.1–2%)

Best For: Sensitive/eczema-prone skin, irritation relief.
What It Does: Soothes with beta-glucans and avenanthramides, reduces itching.
Effectiveness: Reduces inflammation/itching in days; benefits accrue over weeks.
Pair With: Ceramides, Beta-Glucan, Hyaluronic Acid.
Avoid Using With: Oat allergies (rare).
Notes: Potent for eczema; ideal in therapeutic formulations.

Azelaic Acid (Multitasking Acid, 8–20%)

Best For: Sensitive/rosacea-prone skin, acne, PIH.
What It Does: Reduces inflammation, kills bacteria, brightens, exfoliates gently.
Effectiveness: Improves redness/acne in 4–12 weeks; pregnancy-safe.
Pair With: Niacinamide, Hyaluronic Acid, Gentle Retinoids (alternate).
Avoid Using With: High acids (same step).
Notes: Versatile; calms redness while addressing acne.

Bisabolol (Chamomile-Derived Anti-Inflammatory, 0.1–1%)

Best For: Redness, irritation, sensitive skin.
What It Does: Reduces inflammation, soothes, provides antimicrobial benefits.
Effectiveness: Reduces redness in days; improves sensitivity in 2 weeks.
Pair With: Panthenol, Niacinamide, Glycerin.
Avoid Using With: High alcohol, strong astringents (counteract soothing).
Notes: Gentle, non-sensitizing; ideal for reactive skin.

Calendula Officinalis Flower Extract (Soothing Botanical, 2–5%)

Best For: Minor wounds, redness, inflammatory conditions.
What It Does: Calms irritation, supports healing, provides antiseptic benefits.
Effectiveness: Soothes within 24 hours; healing improves over weeks.
Pair With: Aloe Vera, Zinc Oxide, Panthenol.
Avoid Using With: Strong acids, high Vitamin C (may reduce efficacy).
Notes: Gentle; effective for minor skin disruptions.

Centella Asiatica Extract (Gotu Kola) (Healing Botanical, ≥2% Triterpenes)

Best For: Inflamed skin, wound healing, redness reduction.
What It Does: Reduces inflammation, boosts collagen, strengthens barrier.
Effectiveness: Reduces redness in 2 weeks; improves resilience in 4–6 weeks.
Pair With: Niacinamide, Hyaluronic Acid, Madecassoside.
Avoid Using With: High L-Ascorbic Acid (may destabilize).
Notes: Potent for barrier repair; popular in cica products.

Niacinamide (Vitamin B3, 2–10%)

Best For: Sensitive/acne-prone skin, barrier repair, redness.
What It Does: Reduces inflammation, strengthens barrier, regulates oil.
Effectiveness: Improves texture/redness in 4–8 weeks; long-term benefits.
Pair With: Ceramides, Hyaluronic Acid, Azelaic Acid.
Avoid Using With: High L-Ascorbic Acid (same step, if flushing).
Notes: Versatile; ideal for sensitive and combination skin.

Panthenol (Provitamin B5) **(Humectant, 2–5%)**

Best For: Irritation, wound healing, hydration.
What It Does: Hydrates, reduces inflammation, supports barrier repair.

Effectiveness: Improves hydration by 30% in days; resilience in 1–2 weeks.
Pair With: Ceramides, Niacinamide, Allantoin.
Avoid Using With: Strongly acidic formulations (may destabilize).
Notes: Broadly compatible; excellent for post-procedure care.

Tips for Choosing Soothing and Anti-Inflammatory Ingredients

- **Sensitive Skin**: Use allantoin, bisabolol, aloe vera for gentle soothing.
- **Reactive/Redness-Prone Skin**: Choose centella asiatica, niacinamide, azelaic acid.
- **Eczema/Itchy Skin**: Opt for avena sativa extract, panthenol.
- **Post-Procedure**: Prioritize Allantoin, Calendula, Panthenol.
- **Hydration Support**: Pair with hyaluronic acid, glycerin to enhance comfort.
- **Patch-Test**: Test new ingredients, especially botanicals, for sensitivity.

PEPTIDES

Peptides are one of the most rapidly evolving skincare ingredient categories. Peptides are short chains of amino acids that act as cellular messengers that instruct skin cells to perform specific functions. Peptides can target skin concerns with remarkable precision, like stimulating collagen production, inhibiting the muscle contractions that cause wrinkles, fading unwanted pigmentation, and more.

What makes peptides especially exciting is the pace of innovation. Today's biotechnology is rapidly developing and delivering new peptide molecules, each engineered for greater stability, precision, and effectiveness. Innovations now include more stable formulations, enhanced penetration systems, and novel sequences that target previously unreachable cellular pathways. The latest frontier includes biomimetic peptides, designed to mimic naturally occurring peptides found in healthy, youthful skin that allow for more effective cellular communication and repair.

The versatility of peptides allows them to be categorized by their primary functions and target concerns:

- **Skin barrier repair peptides** support recovery and strengthen the protective barrier, making them ideal for sensitive or compromised skin.

- **Skin brightening and pigmentation peptides** provide a gentler alternative to traditional lightening agents by inhibiting melanin production and promoting a more even tone.

- **Collagen and tissue repair peptides** stimulate production of the skin's structural proteins to improve volume, texture, and resilience.

- **Wrinkle-relaxing peptides** help soften expression lines and dynamic wrinkles caused by repeated facial movements like smiling, frowning, or squinting.

- **Hair and lash growth peptides** stimulate follicle activity and enhance keratin production for thicker, longer hair.

Whether you're targeting signs of aging, strengthening your barrier, or seeking a more radiant complexion, there's likely a peptide formulation tailored to your needs. The key is understanding which peptides align with your skin goals and incorporating them thoughtfully into your routine for optimal results.

Skin Barrier & Anti-Inflammatory Peptides

Your skin barrier is your first line of defense against environmental stressors, pollutants, contagions, and moisture loss. When this protective barrier becomes compromised from over-exfoliation, harsh ingredients, environmental damage, or natural aging, skin can become sensitive, irritated, and prone to dehydration. Skin barrier repair peptides work to restore and strengthen this protective layer by stimulating the production of key structural proteins and lipids that maintain barrier integrity.

These specialized peptides help accelerate the skin's natural healing process by promoting the production of ceramides, cholesterol, and fatty acids that form the lipid matrix and essential barrier proteins that are vital for maintaining healthy skin structure. By reinforcing the skin's natural defenses, barrier repair peptides can reduce sensitivity, improve moisture retention, and create a more resilient complexion that's better equipped to handle daily environmental challenges.

Acetyl Dipeptide-1 Cetyl Ester (Calmosensine®) (Soothing Peptide, 0.01–0.1%)

Best For: Sensitive skin, irritation reduction, barrier repair.
What It Does: Stimulates pro-endorphins to soothe discomfort, enhances lipid production for barrier strength.
Effectiveness: Reduces irritation in 2–4 weeks; improves barrier in 6–8 weeks with daily AM/PM use.
Pair With: Centella Asiatica, Ceramides, Hyaluronic Acid, Niacinamide.
Avoid Using With: Strong Exfoliants, Retinoids (same step, sensitive skin).
Notes: Ideal for calming stress-induced sensitivity; used in post-procedure care.

Acetyl Tetrapeptide-15 (Skinasensyl®) (Neuro-Soothing Peptide, 0.01–0.1%)

Best For: Sensitive skin, neurogenic inflammation, barrier repair.
What It Does: Reduces sensitivity via neuropeptide modulation, strengthens barrier with lipid synthesis.
Effectiveness: Reduces redness in 2–4 weeks; improves barrier in 6–8 weeks with daily use.
Pair With: Ceramides, Niacinamide, Panthenol, Hyaluronic Acid.
Avoid Using With: High Exfoliants, Retinoids (same step, initial phase).
Notes: Effective for reactive skin; common in calming serums.

Acetyl Tetrapeptide-22 (Theraderm®) (Stress-Protective Peptide, 0.01–0.1%)

Best For: Sensitive skin, environmental stress, barrier repair.
What It Does: Boosts heat shock proteins to protect against stressors, enhances lipid synthesis.
Effectiveness: Improves resilience in 4–8 weeks; benefits accrue over 6–12 weeks with daily use.
Pair With: Ceramides, Niacinamide, Hyaluronic Acid, Centella Asiatica.
Avoid Using With: High Exfoliants, Retinoids (same step, initial phase).
Notes: Unique for stress protection; suits environmentally damaged skin.

Acetyl Tetrapeptide-40 (Anti-Inflammatory Peptide, 0.05–0.5%)

Best For: Sensitive skin, inflammation, barrier repair.
What It Does: Modulates TRPV1 to calm stinging/itching, supports barrier repair.
Effectiveness: Reduces inflammation in 3–6 weeks with daily use.
Pair With: Centella Asiatica, Ceramides, Panthenol.
Avoid Using With: Strong Exfoliants (may trigger inflammation).
Notes: Ideal for chronic sensitivity; popular in specialized formulations.

Copper Tripeptide-1 (GHK-Cu) (Regenerative Peptide, 0.01–0.1%)

Best For: Aging skin, wound healing, acne scarring, post-procedure.
What It Does: Complex of copper ions that naturally occurs in human plasma and is manufactured for skincare; stimulates collagen and elastin production, promotes wound healing, acts as an antioxidant, and supports skin repair by targeting fibroblasts in the dermis and keratinocytes in the stratum basale.
Effectiveness: Visible repair in 4–8 weeks; long-term benefits with daily use.
Pair With: Hyaluronic Acid, Vitamin C (separate steps), Sunscreen.
Avoid Using With: Strong Acids, Exfoliants (same step, irritation).
Notes: Blue color; store away from light; patch-test for sensitivity.

Palmitoyl Pentapeptide-4 (Matrixyl®) (Signal Peptide, ~2–5% Solution)

Best For: Anti-aging, barrier repair, collagen synthesis.
What It Does: Stimulates collagen/elastin/hyaluronic acid, supports barrier function.
Effectiveness: Improves texture/barrier in 6–12 weeks with daily use.
Pair With: Palmitoyl Tetrapeptide-7, Ceramides, Hyaluronic Acid.
Avoid Using With: Strong Acids, Retinoids (same step, initially).
Notes: Gold standard anti-aging peptide with barrier benefits.

Palmitoyl Tetrapeptide-7 (Anti-Inflammatory Peptide, 0.01–0.1%)

Best For: Inflammation reduction, barrier repair, anti-aging.
What It Does: Reduces inflammation, supports barrier via protein synthesis.
Effectiveness: Reduces redness in 4–8 weeks with daily use.
Pair With: Palmitoyl Pentapeptide-4, Ceramides, Centella Asiatica.
Avoid Using With: Strong Exfoliants (may disrupt calming effects).
Notes: Synergizes with other peptides; ideal for sensitive skin.

Palmitoyl Tripeptide-1 (Signal Peptide, 0.01–0.1%)

Best For: Collagen synthesis, barrier repair, skin firmness.
What It Does: Boosts collagen/elastin, enhances barrier function and firmness.
Effectiveness: Improves texture/barrier in 6–10 weeks with daily use.
Pair With: Ceramides, Hyaluronic Acid, Niacinamide.
Avoid Using With: Strong Exfoliants (initially, to avoid sensitivity).
Notes: Effective for mature skin; supports structural repair.

Pentapeptide-13 (Barrier-Supporting Peptide, 0.01–0.1%)

Best For: Sensitive skin, barrier repair, inflammation reduction.
What It Does: Promotes protein/lipid synthesis, calms irritation, enhances hydration.
Effectiveness: Improves barrier/sensitivity in 4–8 weeks with daily use.
Pair With: Ceramides, Hyaluronic Acid, Niacinamide, Centella Asiatica.
Avoid Using With: High Exfoliants, Retinoids (same step, initial phase).
Notes: Emerging in K-beauty; suits compromised skin.

Tips for Choosing Skin Barrier and Anti-Inflammatory Peptides

- **Sensitive Skin**: Use Acetyl Tetrapeptide-15, Acetyl Dipeptide-1 Cetyl Ester for gentle soothing.
- **Compromised Skin Barrier**: Choose Pentapeptide-13, Palmitoyl Tripeptide-1, Copper Tripeptide-1.
- **Anti-Aging with Skin Barrier Support**: Opt for Palmitoyl Pentapeptide-4, Palmitoyl Tripeptide-1.
- **Redness/Stress**: Prioritize Acetyl Tetrapeptide-22, Acetyl Tetrapeptide-40.
- **Hydration Support**: Pair with Hyaluronic Acid, Ceramides to enhance barrier repair.
- **Patch-Test**: Always patch test peptides, especially Copper Tripeptide-1, for sensitivity.

Skin Brightening & Pigmentation Peptides

Uneven skin tone and hyperpigmentation are among the most common skincare concerns, often resulting from sun damage, hormonal changes, acne scarring, or natural aging. Traditional brightening ingredients like hydroquinone and high-concentration acids can be effective but may cause irritation, especially for sensitive skin, and are associated with various health risks.

Skin brightening peptides provide a gentler yet powerful alternative, working at the cellular level to regulate melanin production and promote a more uniform complexion. These innovative peptides target multiple pathways in the pigmentation process. Some inhibit tyrosinase, the key enzyme responsible for melanin synthesis, while others accelerate cellular turnover to fade existing dark spots more efficiently.

Advanced brightening peptides can also help prevent future pigmentation by strengthening the skin's defenses against UV-induced damage. Unlike harsh bleaching agents, peptide-based treatments work gradually and safely, making them suitable for long-term use and for all skin types, including those prone to sensitivity or irritation.

Acetyl Glycyl Beta-Alanine (GenoWhite®) (Brightening Peptide, 0.1–2%)

Best For: Hyperpigmentation, uneven tone, barrier repair.
What It Does: Inhibits tyrosinase to reduce melanin, enhances barrier with cellular cohesion.
Effectiveness: Reduces pigmentation in 4–8 weeks; improves brightness/barrier in 6–12 weeks with daily AM/PM use.
Pair With: Niacinamide, Hyaluronic Acid, Ceramides, Centella Asiatica.
Avoid Using With: High Exfoliants, Retinoids (same step, initial phase).
Notes: Gentle, K-beauty favorite; store in airtight, light-protected packaging.

Decapeptide-12 (Lumixyl™) (Signal Peptide, 0.1–1%)

Best For: Hyperpigmentation, melasma, dark spots, sensitive skin.
What It Does: Inhibits tyrosinase to fade dark spots, promotes even tone.
Effectiveness: Visible brightening in 8–12 weeks with daily AM use.
Pair With: Vitamin C, Niacinamide, Alpha Arbutin, Sunscreen.
Avoid Using With: High AHAs (same step), extreme pH formulations.
Notes: Gentle alternative to hydroquinone; ideal for sensitive skin.

Hexapeptide-2 (Dermostatyl™ IS) (Signal Peptide, 0.01–0.1%)

Best For: Uneven tone, dark spots, brightening.
What It Does: Inhibits tyrosinase to reduce melanin, promotes even complexion.
Effectiveness: Reduces pigmentation in 6–8 weeks with daily AM use.
Pair With: Nonapeptide-1, Vitamin C, Kojic Acid.
Avoid Using With: Strong Exfoliants (same step, sensitivity).
Notes: Popular in K-beauty serums for targeted brightening.

Nonapeptide-1 (Melanostatine 5) (Melanin-Inhibiting Peptide, 0.01–0.1%)

Best For: Hyperpigmentation, melasma, barrier repair.
What It Does: Blocks α-MSH to reduce melanin, supports barrier cohesion.
Effectiveness: Reduces pigmentation in 4–8 weeks; improves barrier in 6–12 weeks with daily AM/PM use.
Pair With: Niacinamide, Hyaluronic Acid, Ceramides, Sunscreen.
Avoid Using With: High Exfoliants, Retinoids (same step, initial phase).
Notes: Effective for spot treatments; gentle for all skin tones.

Oligopeptide-34 (Brightening Peptide, 0.01–0.1%)

Best For: Hyperpigmentation, PIH, melasma, barrier repair.
What It Does: Inhibits tyrosinase and α-MSH, enhances barrier function.

Effectiveness: Reduces pigmentation in 4–8 weeks; improves barrier in 6–12 weeks with daily AM/PM use.
Pair With: Niacinamide, Hyaluronic Acid, Ceramides, Sunscreen.
Avoid Using With: High Exfoliants, Retinoids (same step, initial phase).
Notes: K-beauty staple; suits sensitive skin, store airtight.

Oligopeptide-58 (Brightening Peptide, 0.01–0.1%)

Best For: Hyperpigmentation, PIH, melasma, barrier repair.
What It Does: Inhibits tyrosinase and α-MSH, strengthens barrier matrix.
Effectiveness: Reduces pigmentation in 4–8 weeks; improves barrier in 6–12 weeks with daily AM/PM use.
Pair With: Niacinamide, Hyaluronic Acid, Alpha Arbutin, Sunscreen.
Avoid Using With: High Exfoliants, Retinoids (same step, initial phase).
Notes: Gentle for sensitive skin; ideal for PIH, store airtight.

Tetrapeptide-21 (Brightening Peptide, 0.01–0.1%)

Best For: Hyperpigmentation, PIH, anti-aging, barrier repair.
What It Does: Inhibits tyrosinase, boosts collagen/elastin for brightness and resilience.
Effectiveness: Reduces pigmentation in 4–8 weeks; improves firmness in 6–12 weeks with daily AM/PM use.
Pair With: Vitamin C, Niacinamide, Hyaluronic Acid, Sunscreen.
Avoid Using With: High Exfoliants, Retinoids (same step, initial phase).
Notes: Dual-action for pigmentation and aging; store in light-protected packaging.

Tips for Choosing Skin Brightening Peptides

- **Sensitive Skin**: Use Decapeptide-12, Nonapeptide-1 for gentle brightening.
- **Melasma/PIH**: Choose Oligopeptide-34, Oligopeptide-58, Tetrapeptide-21.
- **Dark Spots**: Opt for Acetyl Glycyl Beta-Alanine, Hexapeptide-2.
- **Anti-Aging with Brightening**: Prioritize Tetrapeptide-21 for collagen benefits.
- **Sun Protection**: Always pair with Broad-Spectrum Sunscreen to prevent new pigmentation.
- **Patch-Test**: Always patch test peptides for sensitivity, especially with other actives.

Collagen & Tissue Repair Peptides

Collagen is the most abundant protein in our skin. It provides the structural framework that keeps skin firm, plump, and youthful. As we age, natural collagen production begins declining in our 20s, leading to fine lines, wrinkles, and loose, crepey skin over the decades. Environmental factors such as UV exposure, pollution, and lifestyle choices can accelerate this breakdown. This makes collagen support a cornerstone of effective anti-aging skincare.

Collagen and tissue repair peptides work by signaling fibroblasts, the cells that are responsible for producing collagen, elastin, and other structural proteins. These peptides essentially "trick" the skin into initiating repair processes, boosting the production of these key proteins. Some peptides contain fragments of broken-down collagen that act as messengers, while others are designed to mimic growth factors that naturally promote tissue regeneration.

The result is firmer, more resilient skin with improved texture, fewer fine lines, and enhanced structural integrity. Unlike topical collagen that has molecules too large to penetrate the skin effectively, peptides work from within to support your skin's natural repair and rebuilding mechanisms.

Acetyl Tetrapeptide-9 (Structural Support Peptide, 0.1–1%)

Best For: Skin firmness, barrier repair, anti-aging.
What It Does: Boosts collagen and fibronectin synthesis; strengthens skin architecture.
Effectiveness: Improves texture/firmness in 6–10 weeks with daily use.
Pair With: Hyaluronic Acid, Niacinamide, Ceramides.
Avoid Using With: Strong Exfoliants (same step, sensitivity).
Notes: Ideal for mature skin needing structural support.

Acetyl Tetrapeptide-11 (Collagen Synthesis Peptide, 0.01–0.1%)

Best For: Skin firmness, collagen stimulation, anti-aging.
What It Does: Stimulates fibroblast activity for collagen/elastin production, enhances firmness.
Effectiveness: Improves firmness/texture in 6–10 weeks with daily AM/PM use.
Pair With: Hyaluronic Acid, Vitamin C, Ceramides.
Avoid Using With: Strong Acids, Retinoids (initially, irritation).
Notes: Effective for mature skin requiring collagen support.

Copper Tripeptide-1 (GHK-Cu) (Regenerative Peptide, 0.01–0.1%)

Best For: Aging skin, wound healing, acne scarring, post-procedure.
What It Does: Complex of copper ions that naturally occurs in human plasma and is manufactured for skincare; stimulates collagen and elastin production, promotes wound healing, acts as an antioxidant, and supports skin repair by targeting fibroblasts in the dermis and keratinocytes in the stratum basale.
Effectiveness: Visible repair in 4–8 weeks; long-term benefits with daily use.
Pair With: Hyaluronic Acid, Vitamin C (separate steps), Sunscreen.
Avoid Using With: Strong Acids, Exfoliants (same step, irritation).
Notes: Blue color; store away from light; patch-test for sensitivity.

Matrixyl™ 3000 (Peptide Complex, 3–8%)

Best For: Anti-aging, wrinkle reduction, collagen stimulation.
What It Does: Combines Palmitoyl Tripeptide-1 and Palmitoyl Tetrapeptide-7 to boost collagen, reduce inflammation.

Effectiveness: Reduces wrinkles, improves firmness in 6–12 weeks with daily AM/PM use.
Pair With: Hyaluronic Acid, Vitamin C, Niacinamide.
Avoid Using With: High Exfoliants, Retinoids (same step, initially).
Notes: Gold standard for anti-aging; widely studied.

Matrixyl™ Synthe'6 (Peptide Complex, 2–5%)

Best For: Wrinkle reduction, matrix repair, skin smoothing.
What It Does: Contains Palmitoyl Tripeptide-38 to stimulate six matrix components (collagen, fibronectin, etc.).
Effectiveness: Reduces wrinkles, improves smoothness in 8–16 weeks with daily use.
Pair With: Hyaluronic Acid, Vitamin C, Ceramides.
Avoid Using With: Strong Exfoliants (may disrupt matrix repair).
Notes: Advanced Matrixyl for comprehensive anti-aging.

Neodermyl™ (Peptide Complex, 0.5–2%)

Best For: Mature skin, sagging, wrinkle reduction.
What It Does: Combines methylglucoside phosphate and copper/zinc lysinates to boost collagen/elastin, re-energizes fibroblasts.
Effectiveness: Improves firmness in 2–4 weeks; up to 20% wrinkle reduction in 8–12 weeks.
Pair With: *Matrixyl™, Hyaluronic Acid, Vitamin C.
Avoid Using With: Strong Acids, high pH formulations (>8).
Notes: Fast acting; ideal for premium anti-aging serums.

Palmitoyl Pentapeptide-4 (Matrixyl®) (Signal Peptide, ~2–5%)

Best For: Anti-aging, collagen synthesis, skin rejuvenation.
What It Does: Stimulates collagen, elastin, hyaluronic acid, supports barrier function.
Effectiveness: Reduces wrinkles, improves firmness in 6–12 weeks with daily use.
Pair With: Palmitoyl Tetrapeptide-7, Ceramides, Hyaluronic Acid.
Avoid Using With: Strong Acids, Retinoids (initially, irritation).
Notes: Gold standard peptide; extensively researched.

Palmitoyl Tripeptide-1 (Collagen Signal Peptide, 2–5%)

Best For: Collagen stimulation, skin firmness, anti-aging.
What It Does: Boosts collagen/elastin/glycosaminoglycans, improves structural integrity.
Effectiveness: Enhances firmness/texture in 8–12 weeks with daily AM/PM use.
Pair With: Palmitoyl Tetrapeptide-7, Hyaluronic Acid, Vitamin C.
Avoid Using With: Strong Acids, Exfoliants (initially, irritation).
Notes: Key in Matrixyl 3000; synergistic with other peptides.

Tripeptide-10 Citrulline (Decorinyl®) (Collagen Organization Peptide, 0.1–2%)

Best For: Skin firmness, collagen organization, anti-aging.

What It Does: Regulates collagen fiber organization, improves firmness and texture.
Effectiveness: Enhances firmness in 6–10 weeks with daily use.
Pair With: Hyaluronic Acid, Niacinamide, Collagen Peptides.
Avoid Using With: Strong Exfoliants (may disrupt collagen organization).
Notes: Unique for collagen quality; suits mature skin.

Tips for Choosing Collagen and Tissue Repair Ingredients

- **Mature Skin**: Use Matrixyl™ 3000, Matrixyl™ Synthe'6, Neodermyl™ for comprehensive anti-aging.
- **Wound Healing**: Choose Copper Tripeptide-1 for regeneration and scarring.
- **Skin Firmness**: Opt for Acetyl Tetrapeptide-9, Palmitoyl Tripeptide-1, Tripeptide-10 Citrulline.
- **Sensitive Skin**: Prioritize Palmitoyl Pentapeptide-4, Acetyl Tetrapeptide-11 for gentle action.
- **Hydration Support**: Pair with Hyaluronic Acid, Ceramides to enhance repair.
- **Patch-Test**: Especially Copper Tripeptide-1 for sensitivity, especially with actives.

Wrinkle Relaxing & Eye Peptides

Dynamic wrinkles - those lines that form from repeated facial expressions like smiling, frowning, squinting, and raising eyebrows - are among the first visible signs of aging. Unlike deeper static wrinkles caused by collagen loss, dynamic wrinkles result from the constant contraction of facial muscles beneath the skin. Over time, these repetitive movements create permanent creases that remain visible even when the face is at rest.

Wrinkle-relaxing peptides provide a topical alternative to injectable treatments by gently reducing the muscle contractions that contribute to expression lines. These sophisticated peptides target the communication pathway between nerves and muscles, helping to soften facial movements without completely inhibiting natural expression. While they don't smooth wrinkles as quickly or dramatically as injectable neurotoxins like Botox or Dysport, they can deliver visible results over time.

Some peptides work by blocking calcium channels that trigger muscle contractions, while others interfere with the release of neurotransmitters that signal muscles to contract. The result is a gradual smoothing of forehead lines, crow's feet, frown lines, and other expression-related wrinkles, creating a more relaxed, youthful appearance while preserving natural facial mobility.

Because your eye area is one of the most expressive regions of the face, many wrinkle-relaxing peptides are also used in eye formulations. In addition to reducing crow's feet and under eye crinkles, they help address puffiness and dark circles by supporting microcirculation, calming inflammation, and strengthening the skin's structural proteins.

This overlap makes peptides like Argireline®, SNAP-8®, and Leuphasyl valuable for both smoothing dynamic wrinkles and improving delicate, fatigue-prone skin around the eyes, providing a well-rounded, non-invasive approach to maintaining a refreshed, smoother appearance.

Acetyl Hexapeptide-30 (Inyline®) (Neuromuscular Peptide, 0.01–0.1%)

Best For: Deep expression lines, forehead wrinkles, crow's feet.
What It Does: Inhibits SNARE complex to block neurotransmitter release, smooths dynamic wrinkles.
Effectiveness: Reduces wrinkles in 6–12 weeks with daily PM use.
Pair With: Hyaluronic Acid, Vitamin C, *Matrixyl™, Sunscreen.
Avoid Using With: Strong Acids, Alcohol-Based Products (may degrade peptide).
Notes: Potent for deep wrinkles; best in evening routines.

Acetyl Hexapeptide-38 (Adifyline®) (Neuromuscular Peptide, 0.01–0.1%)

Best For: Expression lines, under-eye hollows, volume loss.
What It Does: Inhibits muscle contractions, promotes adipogenesis for volume and wrinkle reduction.
Effectiveness: Improves wrinkles/volume in 4–8 weeks with daily AM/PM use.
Pair With: Hyaluronic Acid, *Matrixyl™, Caffeine, Niacinamide.
Avoid Using With: Strong Acids, Retinoids (same step, initially).
Notes: Unique for volume enhancement; ideal for eye contours.

Acetyl Hexapeptide-8 (Argireline®) (Neuromuscular Peptide, 5–10%)

Best For: Expression lines, crow's feet, forehead wrinkles.
What It Does: Inhibits neurotransmitter release for mild muscle relaxation, smooths dynamic wrinkles.
Effectiveness: Reduces wrinkle depth in 4–8 weeks with daily AM/PM use.
Pair With: *Matrixyl™, Hyaluronic Acid, Niacinamide, Vitamin C.
Avoid Using With: Extreme pH formulations (<3.5 or >8), Acids (same step).
Notes: Popular "topical Botox" alternative; requires consistent use.

Acetyl Octapeptide-3 (SNAP-8®) (Neurotransmitter-Inhibiting Peptide, 3–10%)

Best For: Deep wrinkles, expression lines, forehead treatments.
What It Does: Inhibits neurotransmitter release, smooths deep expression lines.
Effectiveness: Reduces wrinkle depth in 4–8 weeks with daily use.
Pair With: *Matrixyl™, Hyaluronic Acid, Ceramides.
Avoid Using With: Strong Exfoliants, Retinoids (same step, sensitivity).
Notes: More potent than Argireline®; ideal for deeper wrinkles.

Acetyl Tetrapeptide-5 (Anti-Edema Peptide, 0.01–0.1%)

Best For: Under-eye puffiness, dark circles, fluid retention.

What It Does: Reduces capillary permeability, enhances lymphatic drainage, inhibits glycation.
Effectiveness: Reduces puffiness in 2–4 weeks with AM/PM use.
Pair With: Caffeine, Vitamin K, Hyaluronic Acid.
Avoid Using With: Strong Acids, heavy oils (may cause milia).
Notes: Best applied with gentle patting; morning use for puffiness.

Dipeptide Diaminobutyroyl Benzylamide Diacetate (SYN®-AKE) (Neurotransmitter-Inhibiting Peptide, 1–4%)

Best For: Expression lines, deep wrinkles, eye area.
What It Does: Mimics snake venom to inhibit muscle contractions, smooths wrinkles.
Effectiveness: Reduces wrinkles in 4–6 weeks with daily use.
Pair With: Hyaluronic Acid, *Matrixyl™, Ceramides.
Avoid Using With: Strong Acids, Retinoids (same step, irritation).
Notes: Fast-acting; popular in targeted wrinkle treatments.

Pentapeptide-18 (Leuphasyl®) (Neurotransmitter-Inhibiting Peptide, 0.01–0.1%)

Best For: Forehead wrinkles, crow's feet, precise wrinkle treatment.
What It Does: Modulates calcium channels to reduce muscle contractions, smooths expression lines.
Effectiveness: Reduces wrinkles in 6–10 weeks with daily PM use.
Pair With: Neuromuscular Peptides, Hyaluronic Acid, Antioxidants.
Avoid Using With: Strong Acids, high pH products (>8).
Notes: Best for targeted areas; apply to clean skin PM.

Tips for Choosing Wrinkle Relaxing & Eye Peptides

- **Expression Lines:** Use Acetyl Hexapeptide-8, Acetyl Octapeptide-3, Dipeptide Diaminobutyroyl Benzylamide Diacetate for dynamic wrinkles.
- **Deep Wrinkles:** Choose Acetyl Hexapeptide-30, Pentapeptide-18 for potent action.
- **Eye Puffiness:** Opt for Acetyl Tetrapeptide-5 with caffeine for drainage.
- **Volume Loss:** Prioritize Acetyl Hexapeptide-38 for hollow under-eyes.
- **Hydration Support:** Pair with Hyaluronic Acid, Ceramides for optimal results.
- **Patch-Test:** Always patch test peptides, especially around eyes, for sensitivity.

Hair & Lash Growth Peptides

Just as peptides can influence skin repair and wrinkle formation, certain peptides support stronger, thicker, fuller, and more resilient hair and lashes. Healthy hair and full lashes are often viewed as signs of vitality, youth, and confidence.

Hair and lash growth peptides work by stimulating activity in the hair follicle, extending the growth phase (anagen) of the hair cycle, and improving the anchoring of the hair shaft. Some also enhance blood flow and nutrient delivery to the follicle, creating an optimal environment for stronger, healthier growth.

In lash and brow formulas, peptides are often combined with conditioning agents to reduce breakage and improve density. Many of these formulations also include anti-inflammatory agents, botanical extracts, and vitamins that work synergistically to improve follicle health and support more robust growth cycles.

Hair and lash peptides are not miracle cures for advanced hair loss and are not likely to be ineffective when hair loss is due to underlying medical conditions. However, they can play a meaningful role in maintaining fullness and preventing thinning when used consistently. With their ability to fortify follicles and extend growth cycles, they represent one of the most promising frontiers in peptide-based skincare and haircare.

Biotinoyl Tripeptide-1 (Hair Strengthening Peptide, 1–5%)

Best For: Thinning hair, sparse lashes/brows, follicle strengthening.
What It Does: Stimulates collagen IV/laminin in follicles, reduces hair loss, enhances thickness.
Effectiveness: Improves density, lash/brow thickness in 8–16 weeks with daily use.
Pair With: Panthenol, Copper Peptides, Biotin, Caffeine.
Avoid Using With: Extreme pH formulations (<3.5 or >8).
Notes: Biotin enhances penetration; ideal for lash/brow serums.

Myristoyl Pentapeptide-17 (Lash Growth Peptide, 0.1–0.5%)

Best For: Sparse lashes, lash breakage, brow thinning.
What It Does: Stimulates keratin synthesis, extends anagen phase for lash growth.
Effectiveness: Enhances lash length/volume in 8–16 weeks with daily PM use.
Pair With: Biotin, Panthenol, Hyaluronic Acid.
Avoid Using With: Oil-based Makeup Removers, harsh rubbing.
Notes: Professional grade ingredient; apply to clean lash line PM.

Palmitoyl Tetrapeptide-20 (Greyverse™) (Hair Pigmentation Peptide, 1–5%)

Best For: Premature graying, hair color preservation.
What It Does: Stimulates melanocyte activity to maintain hair pigment, slows graying.
Effectiveness: Reduces gray hair percentage in 12–24 weeks with daily use.
Pair With: Biotin, Copper Peptides, Panthenol, Antioxidants.
Avoid Using With: Extreme pH formulations (<3.5 or >8).
Notes: Patience required; best for preventing further graying.

---

sh-Oligopeptide-2 (IGF-1) (Growth Factor Peptide, 0.1–1 ppm)

Best For: Thinning hair, sparse lashes, follicle health.
What It Does: Enhances follicle stem cell activity, extends anagen phase, boosts density.
Effectiveness: Improves hair/lash density in 8–12 weeks with daily AM/PM use.
Pair With: Biotinoyl Tripeptide-1, Panthenol, DHT Blockers.
Avoid Using With: Strong Acids, Alcohol-Based Products (may denature peptide).

Notes: Bioengineered for premium hair/lash growth serums.

Tips for Choosing Hair & Lash Growth Peptides

- **Thinning Hair:** Choose Biotinoyl Tripeptide-1 for follicle regeneration.
- **Sparse Lashes/Brows:** Choose Myristoyl Pentapeptide-17, sh-Oligopeptide-2 for growth.
- **Premature Graying:** Opt for Palmitoyl Tetrapeptide-20 to preserve color.
- **Scalp Health:** Pair with Panthenol, Caffeine for nourishment and circulation.
- **Application:** Apply lash peptides to clean lash line PM; avoid oils post-application.
- **Patch-Test:** Always patch test for sensitivity, especially for lash products near eyes.

ADVANCED REGENERATIVE BIOTECHNOLOGY

The modern era of regenerative skincare began in 2001 when SkinMedica launched TNS Recovery Complex, the first widely available product featuring human-derived topical growth factors. Since then, our understanding of cellular communication has accelerated, and with it the sophistication of skincare actives that tap the skin's innate repair programs.

Over the 2010s, formulations evolved from simple peptides into complex systems incorporating bioengineered growth factors, conditioned media from human cell cultures, and early stem cell–inspired technologies. In parallel, brands introduced plant stem cell extracts, not live cells, but cultured tissue extracts rich in protective phytonutrients from sources like Swiss apple, edelweiss, and grape.

By the mid-2010s, extracellular vesicles (EVs), including exosomes and PDRN (polydeoxyribonucleotide) began appearing in clinical aesthetics and Korean (K-beauty) products, followed more recently by plant-derived vesicles (often marketed as exosome-like nanoparticles). Together, these classes represent a spectrum: from human-derived regenerative messengers that can direct repair, to plant-derived technologies that emphasize antioxidant delivery, barrier support, and signaling analogs with fewer sourcing and regulatory constraints.

This chapter explains what these technologies are, how they work, and where they differ. You'll learn why source, delivery, stability, and dose determine outcomes; how professional procedures (e.g., microneedling, microchanneling, fractional lasers) can improve penetration for large or fragile active. Covered are:

- **Bioengineered Growth Factors:** Lab-made replicas of the body's own repair proteins that signal fibroblasts and keratinocytes to stimulate collagen, elastin, and cellular renewal.

- **Conditioned Media (Stem Cell–Derived Actives):** Nutrient-rich fluid collected from cultured cells that delivers a mix of growth factors and cytokines that support skin regeneration, wound healing, and tissue repair.

- **Extracellular Vesicles:** Tiny membrane-bound messengers like exosomes and platelet-derived vesicles that deliver concentrated regenerative signals to calm inflammation and accelerate recovery after procedures.

- **Plant-Derived Nanovesicles & Stem Cell Extracts:** Vesicles and extracts sourced from plants that provide antioxidants, phytonutrients, and protective signals to support barrier strength and defend against environmental stress.

- **PDRN & Polynucleotides:** DNA fragments derived from salmon or bioengineered sources that reduce inflammation, improve tissue repair, and promote collagen production by activating cellular repair pathways.

This chapter clarifies what each technology can and cannot do by outlining mechanisms so you can make better informed decisions about incorporating advanced biotechnology into modern skincare formulations and treatment protocols.

Bioengineered Growth Factors & Exosomes

Growth factors are naturally occurring proteins that act as the body's master messengers for healing and regeneration. In skin, they play a central role in cellular communication, telling fibroblasts to produce collagen, signaling keratinocytes to renew, and directing tissue repair after injury. Without growth factors, wound healing and everyday skin maintenance would grind to a halt.

Unlike many skincare ingredients that work on the surface, growth factors can influence activity deeper within the skin, where true regeneration occurs. However, their effectiveness depends heavily on the source quality, delivery method, and formulation integrity.

Growth factors are large protein molecules, which makes it difficult for them to penetrate the skin barrier effectively. For optimal effectiveness, they often require a controlled injury to the skin, such as microneedling, microchanneling, or fractional laser treatments, to create physical pathways that allow growth factors to reach target cells in the deeper dermal layers.

These proteins are also fragile and degrade quickly when exposed to light, heat, or oxygen. To be effective, they often require synergistic ingredients like peptides or antioxidants, which adds to manufacturing complexity and cost. Stabilizing these formulas often requires advanced encapsulation or even refrigeration that further increase production and storage costs, and ultimately, consumer price.

Despite challenges in sourcing, delivery, and stability, growth factors remain one of the most advanced and potentially transformative frontiers in regenerative aesthetics.

Understanding Growth Factor Naming Conventions

A naming convention helps identify the technology used to create bioengineered growth factors:

- **"rh-"** stands for recombinant human. This means the growth factor was manufactured with a bioengineering process where a human gene is inserted into a host organism (such as E. coli bacteria or plant cells like tobacco), turning that organism into a factory that produces an exact replica of the desired human peptide. This method is commonly used to create EGF (epidermal growth factor) in skincare.

- **"sh-"** stands for synthetic human. This means the growth factor was created using a synthesized copy of a human gene, not one isolated from actual human cells. The "sh-" prefix guarantees the ingredient is a fully bioengineered, ethically sourced replica. This is important in regions like the EU, where strict regulations govern the use of ingredients derived from human tissue.

While suppliers may use different production methods, the molecules produced (e.g., rh-Oligopeptide-1 and sh-Oligopeptide-1) are chemically identical. For clarity and to avoid duplication, only the synthetic "sh-" versions, which are growing in popularity, are listed in the ingredient details that follow.

---

sh-Oligopeptide-1 (EGF) (Cell Growth Factor, 0.00001–0.1%)

Best For: Mature skin, wound healing, anti-aging, skin regeneration.
What It Does: Mimics human EGF to stimulate keratinocyte/fibroblast proliferation, boosts collagen/elastin, reduces wrinkles.
Effectiveness: Improves texture, firmness in 8–12 weeks with daily AM/PM use.
Pair With: Niacinamide, Hyaluronic Acid, *Matrixyl™, Vitamin C.
Avoid Using With: Strong Acids (<3.5 pH), high pH (>8), harsh exfoliants.
Notes: Potent, bioidentical EGF; premium anti-aging ingredient.

---

sh-Oligopeptide-2 (IGF-1) (Growth Factor Peptide, 0.1–1 ppm)

Best For: Skin firmness, thinning hair, sparse lashes, anti-aging.
What It Does: Mimics IGF-1 to enhance cellular metabolism, collagen synthesis, and follicle growth for skin and hair.
Effectiveness: Improves skin firmness, hair/lash density in 8–12 weeks with daily AM/PM use.
Pair With: sh-Oligopeptide-1, Biotinoyl Tripeptide-1, Hyaluronic Acid.
Avoid Using With: Strong Acids, Alcohol-Based Products, harsh treatments.
Notes: Versatile for skin and hair; emerging in lash serums.

---

sh-Oligopeptide-4 (FGF-2, bFGF) (Fibroblast Growth Factor, 0.1–1 ppm)

Best For: Deep wrinkles, skin aging, tissue repair, post-procedure.
What It Does: Mimics bFGF to promote fibroblast proliferation, angiogenesis, and collagen/elastin synthesis.
Effectiveness: Reduces wrinkles, improves firmness in 2–4 weeks with daily AM/PM use.
Pair With: sh-Oligopeptide-1, *Matrixyl™, Hyaluronic Acid, Vitamin C.

Avoid Using With: Strong Acids, Alcohol-Based Products, harsh exfoliants.
Notes: Potent for premium anti-aging and repair serums.

---

sh-Polypeptide-1 (bFGF) (Fibroblast Growth Factor, 0.1–2 ppm)

Best For: Deep wrinkles, skin aging, intensive repair.
What It Does: Stimulates fibroblast proliferation, angiogenesis, and collagen/elastin for skin renewal.
Effectiveness: Significant improvements in 3–6 weeks with daily AM/PM use.
Pair With: Growth Factors, Peptide Complexes, Antioxidants.
Avoid Using With: Strong Exfoliants, Alcohol-Based Products, harsh treatments.
Notes: Professional grade; best with expert guidance.

---

sh-Polypeptide-3 (PDGF) (Platelet-Derived Growth Factor, 0.1–1 ppm)

Best For: Wound healing, tissue repair, post-procedure recovery.
What It Does: Stimulates fibroblast migration, accelerates healing, enhances tissue repair.
Effectiveness: Healing improvements in 1–2 weeks; optimal results in 2–4 weeks with daily use.
Pair With: Healing Growth Factors, Anti-Inflammatory Peptides, Moisturizers.
Avoid Using With: Strong Acids, pro-inflammatory ingredients.
Notes: Ideal for post-procedure; requires professional timing.

---

sh-Polypeptide-7 (HGF) (Hepatocyte Growth Factor, 0.1–1 ppm)

Best For: Skin revitalization, tone improvement, post-procedure.
What It Does: Promotes cellular proliferation, microcirculation, and tissue repair for radiance.
Effectiveness: Enhances radiance, texture in 3–6 weeks with daily AM/PM use.
Pair With: sh-Oligopeptide-1, Hyaluronic Acid, Antioxidants.
Avoid Using With: Strong Acids, Alcohol-Based Products, harsh treatments.
Notes: Emerging in serums for vitality; premium ingredient.

---

sh-Polypeptide-10 (KGF, FGF-7) (Keratinocyte Growth Factor, 0.1–1 ppm)

Best For: Epidermal repair, sensitive skin, barrier restoration.
What It Does: Stimulates keratinocyte proliferation, strengthens skin barrier.
Effectiveness: Improves barrier function in 2–4 weeks with daily use.
Pair With: Ceramides, Hyaluronic Acid, Barrier Peptides.
Avoid Using With: Strong Exfoliants, Alcohol-Based Products.
Notes: Gentle; ideal for compromised or sensitive skin.

---

sh-Polypeptide-13 (Noggin) (Hair Regeneration Growth Factor, 0.1–1 ppm)

Best For: Hair loss, androgenetic alopecia, scalp treatments.
What It Does: Antagonizes BMP signaling to promote hair follicle regeneration, enhances density.

Effectiveness: Improves hair density in 8–16 weeks with daily use.
Pair With: Hair Growth Factors, DHT Blockers, Panthenol.
Avoid Using With: Harsh Scalp Treatments, Alcohol-Based Products.
Notes: For advanced hair restoration; professional use advised.

---

sh-Polypeptide-19 (TGF-β3) (Transforming Growth Factor, 0.1–1 ppm)

Best For: Scar prevention, wound healing, post-acne recovery.
What It Does: Promotes scarless healing, regulates collagen, reduces inflammation.
Effectiveness: Reduces scarring in 2–4 weeks; optimal results in 2–6 months.
Pair With: Healing Growth Factors, Anti-Inflammatory Ingredients.
Avoid Using With: Pro-inflammatory ingredients, harsh treatments.
Notes: Ideal for keloid-prone skin; professional timing advised.

Tips for Choosing Bioengineered Growth Factors

- **Anti-Aging**: Use sh-Oligopeptide-1 (EGF), sh-Oligopeptide-4 (FGF-2) for wrinkle reduction.
- **Wound Healing**: Choose sh-Polypeptide-3 (PDGF), sh-Polypeptide-19 (TGF-β3) for repair.
- **Hair Restoration**: Opt for sh-Polypeptide-13 (Noggin) for follicle regeneration.
- **Sensitive Skin**: Prioritize sh-Polypeptide-10 (KGF) for barrier repair.
- **Hydration Support**: Pair with Hyaluronic Acid, Ceramides for enhanced results.
- **Patch-Test**: Aways patch test for sensitivity, especially post-procedure or on compromised skin.

Human Stem Cell-Derived Actives

Conditioned media refers to the nutrient-rich fluid collected from cultured stem cells, most commonly human fibroblasts or mesenchymal stem cells, after they've secreted a complex mix of growth factors, cytokines, and peptides. When human-derived, these bioactive compounds act as the skin's natural messengers, capable of stimulating repair, enhancing collagen synthesis, and improving overall skin function.

In your body, these signals naturally drive repair, regeneration, and communication between cells. In contrast, plant stem cell–derived conditioned media does not contain human growth factors. However, it delivers antioxidant and protective phyto-molecules that support skin resilience through different biological pathways. The source of conditioned media significantly influences its composition and effectiveness in skincare formulations.

- **Human Fibroblasts**: Sourced from neonatal foreskin or adult skin tissue, human fibroblasts are cultured in controlled environments to produce conditioned media rich in growth factors like epidermal growth factor (EGF), fibroblast growth factor (FGF), and transforming growth factor-beta (TGF-β), along with cytokines and matrix proteins.

- **Bone Marrow Mesenchymal Stem Cells (BM-MSCs):** Extracted from bone marrow, BM-MSCs are multipotent stem cells grown in lab cultures. Their conditioned media is rich in sh-Polypeptide-9 (VEGF) and sh-Polypeptide-149 (IGF-1 variant), which promote angiogenesis, collagen production, and anti-inflammatory responses. Exosomes in BM-MSC media enhance intercellular communication for deeper repair.

- **Adipose Stem Cells (ASCs):** Derived from fat tissue, often via liposuction or donor banks, ASCs produce conditioned media containing sh-Polypeptide-123 (a bFGF variant), anti-inflammatory cytokines, and lipids that support skin barrier repair and stimulate fibroblast activity.

- **Umbilical and Placental Stem Cell Conditioned Media:** This category includes conditioned media from:
  - Placental stem cells (PSC-CM)
  - Umbilical cord mesenchymal stem cells (UC-MSC-CM)
  - Wharton's jelly stem cells (WJ-MSC-CM)

  Sourced ethically from human placental or umbilical tissue, these are rich in sh-Oligopeptide-1 (EGF), sh-Polypeptide-9 (VEGF), sh-Polypeptide-121 (aFGF), and sh-Polypeptide-149 (IGF-1 variant). They also contain exosomes and hyaluronic acid, making them highly effective for tissue remodeling, hydration, and post-procedure recovery.

- **Plant Stem Cells:** Typically sourced from apple (Malus domestica), grape (Vitis vinifera), or edelweiss, plant stem cells yield conditioned media containing antioxidants, flavonoids, and pea peptides. While they lack human growth factors, they help defend against oxidative stress and UV damage by mimicking some protective signaling pathways.

Just as important as the source is how these bioactives are stabilized and delivered. Growth factors and cytokines are fragile proteins that degrade quickly when exposed to light, oxygen, or incorrect pH levels. Without proper stabilization, they may never reach the deeper dermal layers where they're most effective.

To overcome this, manufacturers use encapsulation technologies like:

- Liposomes
- Nanovesicles (exosomes)
- Hydrogel matrices

These delivery systems protect the active ingredients and enhance skin penetration. Some formulations also pair conditioned media with stabilizing peptides or hyaluronic acid to extend activity and improve compatibility with the skin.

Unlike isolated ingredients, conditioned media delivers a dynamic, synergistic blend of regenerative signals that mimic the body's own healing processes. It represents a leap forward in topical rejuvenation, providing visible improvements in texture, tone, and resilience, especially when used alongside professional treatments that improve skin permeability.

While source quality and formulation stability are key, stem cell–derived conditioned media remains one of the most promising innovations for activating true biological renewal.

Human Adipose Derived Stem Cell Conditioned Media (Growth Factor Complex, 1–10%)

Best For: Anti-aging, skin/hair regeneration, wound healing.
What It Does: Delivers growth factors/cytokines for tissue regeneration, reduces inflammation.
Effectiveness: Visible improvements in 2–6 weeks; optimal in 2–4 months with daily use.
Pair With: Hyaluronic Acid, Antioxidants, Microneedling.
Avoid Using With: Strong Acids, Alcohol-Based Products, harsh treatments.
Notes: Potent for skin/hair; requires sterile handling for breached skin.

Human Bone Marrow Stem Cell Conditioned Media (Growth Factor Complex, 1–10%)

Best For: Deep regeneration, anti-aging, professional treatments.
What It Does: Provides growth factors/cytokines for tissue repair, rejuvenation.
Effectiveness: Significant results in 3–6 weeks; optimal in 2–4 months with daily use.
Pair With: Hyaluronic Acid, Peptides, Microneedling.
Avoid Using With: Strong Acids, Alcohol-Based Products, harsh treatments.
Notes: Premium ingredient; requires professional guidance, sterile handling.

Human Fibroblast Conditioned Media (Growth Factor Complex, 1–10%)

Best For: Collagen stimulation, firmness, dermal regeneration.
What It Does: Targets fibroblast activity, boosts collagen, supports dermal repair.
Effectiveness: Enhances firmness in 4–8 weeks; optimal in 2–4 months with daily use.
Pair With: Peptides, Hyaluronic Acid, Vitamin C.
Avoid Using With: Strong Acids, harsh treatments.
Notes: Targets dermal layer; sterile handling for breached skin.

Human Umbilical Mesenchymal Stem Cell Conditioned Media (Growth Factor Complex, 1–10%)

Best For: Intensive regeneration, anti-aging, hair restoration.
What It Does: Provides rich growth factors/exosomes for tissue repair, rejuvenation.
Effectiveness: Visible results in 2–4 weeks; optimal in 1–3 months with daily use.
Pair With: Hyaluronic Acid, Peptides, Microneedling.
Avoid Using With: Strong Acids, Alcohol-Based Products, harsh treatments.
Notes: Exceptional potency; requires ethical sourcing, sterile handling.

Tips for Choosing Human Stem Cell Derived Actives

- **Anti-Aging**: Use Human Placental Stem Cell Exosomes, Human Umbilical Mesenchymal Stem Cell Exosomes for potent rejuvenation.
- **Sensitive Skin**: Choose Human Bone Marrow Stem Cell Exosomes for gentle, anti-inflammatory effects.

- **Wound Healing**: Opt for Human Platelet Extract, Human Adipose Derived Stem Cell Conditioned Media for repair.
- **Collagen Boost**: Prioritize Human Fibroblast Conditioned Media for dermal firmness.
- Professional Use: Pair with Microneedling for enhanced delivery; use sterile products on breached skin.
- **Patch-Test**: Always patch test for sensitivity, especially for post-procedure or compromised skin.

Extracellular Vesicles (Exosomes & PEVs)

Extracellular vesicles (EVs) are membrane-bound nanostructures naturally released by cells. They serve as sophisticated signaling vehicles that transport proteins, RNAs, lipids, and other bioactive molecules between cells, playing a key role in intercellular communication and influencing numerous biological processes.

EVs include exosomes and platelet-derived extracellular vesicles (PEVs), ingredients that represent the most advanced ingredients in regenerative skincare. These microscopic carriers deliver peptides, growth factors, and signaling molecules directly to skin cells, where they powerfully stimulate repair, regulate inflammation, and strengthen the skin's natural defenses.

In the body, EVs act as biological couriers, shuttling instructions between cells to coordinate healing, renewal, and immune responses. In skincare, they provide the potential to deliver regenerative cargo deeper into the skin, where they can stimulate collagen and elastin synthesis, reduce inflammation, and accelerate recovery, particularly when applied after professional treatments such as microneedling, microchanneling, or laser resurfacing.

Unlike whole conditioned media that contains a broad mix of secreted proteins, EVs, particularly exosomes, nanovesicles, and PEVs, provide targeted precision. These refined vesicles deliver stable, intact bioactive payloads with greater potential for penetration and higher specificity, making them especially valuable in:

- Advanced anti-aging serums
- Regenerative facials
- Post-procedure recovery protocols

In skincare applications, exosomes are most often harvested from stem cells or fibroblasts, then purified and concentrated for topical use. Each exosome source carries a unique profile of growth factors, cytokines, and peptides, meaning the cellular origin directly affects the skin benefits provided. However, the use of exosomes in skincare products remains controversial and highly regulated:

- In the United States, exosomes can be used topically but are not FDA-approved for injection or therapeutic use.

- In the European Union, at the time of this writing, exosomes are banned in skincare formulations due to unresolved safety and regulatory concerns.

Ongoing debates focus on manufacturing consistency, purity of preparations, potential immune reactions, and lack of long-term safety data in topical use.

The next leap forward is bioengineered exosomes, lab-created vesicles that mimic the structure and function of natural EVs, but with improved stability, targeted delivery, and customizable payloads. These synthetic vesicles can be fine-tuned to carry:

- Specific peptides
- Antioxidants
- RNA fragments
- Growth factor mimetics

This level of control enables personalized, performance-driven skincare that works in harmony with the skin's own biology.

Exosomes and PEVs represent one of the most exciting frontiers in regenerative skincare, providing unprecedented precision signaling, enhanced delivery, and deep cellular impact. However, their effectiveness, safety, and regulatory status depend heavily on their source, formulation integrity, and method of use.

As research and bioengineering techniques evolve, extracellular vesicles may become central to the future of targeted skin regeneration. For now, their use requires informed selection.

Bioengineered Exosomes (Synthetic Biomimetic, Biosomes) (Cellular Communication Vesicles, Not Specified %)

Best For: Sensitive skin, anti-aging, consistent regenerative results.
What It Does: Lab-created vesicles mimic natural exosomes, delivering standardized growth factors and peptides to promote cellular renewal, collagen synthesis, and repair.
Effectiveness: Improves texture, firmness in 6–8 weeks with daily use.
Pair With: Hyaluronic Acid, Vitamin C, Peptides, Ceramides.
Avoid Using With: Strong Exfoliants, high-potency Retinoids.
Notes: Engineered for predictability; safer alternative to natural exosomes.

Human Adipose Stromal Cell Exosomes (Communication Vesicles, Not Specified %)

Best For: Mature skin, anti-aging, firmness, elasticity.
What It Does: Delivers growth factors from adipose stem cells to boost collagen, enhance repair, and improve skin texture.
Effectiveness: Enhances elasticity, reduces lines in 6–10 weeks with daily use.
Pair With: Hyaluronic Acid, Peptides, Vitamin E, Ceramides.
Avoid Using With: Strong Acids (AHA/BHA), Retinoids, harsh exfoliants.
Notes: Rich in bioactives; regulatory concerns limit use in some regions.

Human Bone Marrow Stem Cell Exosomes (Communication Vesicles, Not Specified %)

Best For: Sensitive/reactive skin, anti-inflammatory, gentle regeneration.
What It Does: Carries anti-inflammatory signals from bone marrow stem cells, calms irritation, supports healing.
Effectiveness: Reduces inflammation, improves repair in 4–6 weeks with daily use.
Pair With: Niacinamide, Hyaluronic Acid, Antioxidants.
Avoid Using With: Strong Acids, Retinoids, irritating actives.
Notes: Gentle; ideal for sensitive skin, but regulatory restrictions apply.

Human Placental Stem Cell Exosomes (Communication Vesicles, Not Specified %)

Best For: Advanced anti-aging, comprehensive skin renewal.
What It Does: Delivers potent growth factors from placental stem cells for cellular repair and vitality.
Effectiveness: Dramatic improvements in texture, firmness in 4–8 weeks with daily use.
Pair With: Hyaluronic Acid, Antioxidants, Peptides.
Avoid Using With: Strong Acids, Retinoids, harsh treatments.
Notes: Highly potent; EU ban and ethical sourcing concerns.

Human Platelet Extract (PEVs) (Platelet-Derived Nanoparticles, Not Specified %)

Best For: Wound healing, skin repair, post-procedure recovery.
What It Does: Delivers concentrated growth factors and anti-inflammatory molecules to accelerate healing and regeneration.
Effectiveness: Faster healing reduced inflammation in 2–4 weeks with daily use.
Pair With: Ceramides, Panthenol, Allantoin, Hyaluronic Acid.
Avoid Using With: Strong Acids, irritating actives.
Notes: Mimics natural healing; regulatory limits for cosmetic use.

Human Umbilical Mesenchymal Stem Cell Exosomes (Communication Vesicles, Not Specified %)

Best For: Premium anti-aging, youthful skin function.
What It Does: Delivers youthful regenerative signals from umbilical cord stem cells, enhances texture and vitality.
Effectiveness: Dramatic improvements in 4–6 weeks with daily use.
Pair With: Hyaluronic Acid, Peptides, Vitamin E.
Avoid Using With: Harsh Exfoliants, Acids, aggressive actives.
Notes: Youthful profile; premium ingredient with regulatory constraints.

Tips for Choosing Extracellular Vesicles (Exosomes & PEVs)

- **Anti-Aging**: Use Human Placental Stem Cell Exosomes, Human Umbilical Mesenchymal Stem Cell Exosomes for potent rejuvenation.
- **Sensitive Skin**: Choose Human Bone Marrow Stem Cell Exosomes for gentle, anti-inflammatory effects.
- **Wound Healing**: Opt for Human Platelet Extract (PEVs) for rapid repair.
- **Consistency:** Prioritize Bioengineered Exosomes for standardized results.

- **Professional Use:** Pair with Microneedling or post-procedure protocols; ensure sterile products.
- **Regulatory Caution:** Check local human derived skincare regulations.

Plant Stem Cell-Derived Actives

Plant-derived nanovesicles and stem cell extracts are emerging as innovative active ingredients in advanced skincare. These botanical technologies provide benefits such as calming inflammation, enhancing hydration, and strengthening the skin barrier, making them a promising alternative to human cell–derived ingredients.

Plant nanovesicles are nanosized, naturally occurring delivery vehicles extracted from whole plant tissues, such as leaves, roots, or fruit. Through a process of blending, filtration, centrifugation, and purification, a concentrated suspension of vesicles is produced. These vesicles carry antioxidants, peptides, and plant-derived signaling molecules to support skin health.

Plant stem cell extracts, on the other hand, are obtained from cultured plant tissues rather than live cells. Much like human cell cultures used to create growth factor and exosome products, plant cultures are sustained in controlled lab environments to yield phytonutrient-rich extracts. These extracts are rich in antioxidants, organic compounds, and protective metabolites that support the skin's resilience.

While plant cells do not produce the same growth factors or cytokines as human cells, and human cells lack receptors to respond to plant-specific signaling, they still deliver valuable bioactivity. Plant nanovesicles mimic some of the communication behavior of human exosomes. Their antioxidant-rich lipids, peptides, and RNAs help reduce inflammation, support repair, and fortify the skin barrier.

Plant stem cell extracts, often derived from meristematic tissue (the regenerative zones of plants), provide compounds that protect against oxidative stress, support collagen synthesis, and enhance elasticity. Common sources include:

- **Malus Domestica:** Apple. rich in protective antioxidants
- **Leontopodium Alpinum:** Edelweiss, known for its resilience in harsh conditions
- **Argania Spinosa:** Argan, supports skin regeneration and elasticity

Although plant-derived actives don't stimulate collagen or keratinocyte activity in the same way human-derived growth factors do, they provide a biocompatible, sustainable, and ethically uncomplicated approach to skin rejuvenation.

With their potent antioxidant profile and growing bioengineering sophistication, plant stem cell–based ingredients represent a next-generation strategy in skincare, merging botanical intelligence with high-performance skin renewal.

Argania Spinosa Callus Culture Extract (Argan Tree Stem Cells) (Plant Stem Cell Extract, 0.5–2%)

Best For: Skin protection, anti-aging prevention, environmental stress defense.
What It Does: Provides antioxidants from argan stem cells to protect against stressors, support repair, and maintain skin vitality.
Effectiveness: Enhances resilience in 4–8 weeks with daily use.
Pair With: Vitamin C, Antioxidants, Moisturizers, Sunscreen.
Avoid Using With: Strong Oxidizing Agents, harsh treatments.
Notes: Preventative; sustainable; ideal for protective routines.

Brassica Oleracea Italica (Broccoli) Vesicles (Plant Extracellular Vesicle Extract, 0.1–1%)

Best For: Antioxidant protection, pollution defense, detoxification.
What It Does: Delivers sulforaphane/glucosinolates via vesicles for cellular protection and detoxification.
Effectiveness: Protective benefits in 2–4 weeks with daily use.
Pair With: Vitamin C, Antioxidants, pollution-defense products.
Avoid Using With: Strong Oxidizing Agents, harsh treatments.
Notes: Advanced vesicle delivery; suits urban environments.

Centella Asiatica Leaf Extract (Anti-Inflammatory Plant Extract, 0.1–5%)

Best For: Sensitive skin, inflammation, wound healing, acne-prone skin.
What It Does: Provides asiaticoside/madecassoside for anti-inflammatory, healing, and barrier support.
Effectiveness: Soothes in 1–3 days; strengthens in 2–6 weeks with daily use.
Pair With: Niacinamide, Hyaluronic Acid, Ceramides.
Avoid Using With: Strong Acids, Exfoliants, Alcohol-Based Products.
Notes: Well-researched; gentle for all skin types.

Cucumis Sativus Fruit Juice Vesicles (Plant Extracellular Vesicle Extract, 0.1–2%)

Best For: Sensitive skin, inflammation, puffiness, hydration.
What It Does: Delivers cucumber-derived vesicles with vitamin C/caffeic acid for soothing and hydration.
Effectiveness: Immediate soothing; benefits in 1–3 days with daily use.
Pair With: Hyaluronic Acid, Aloe Vera, soothing ingredients.
Avoid Using With: Strong Actives, harsh treatments.
Notes: Cooling; ideal for irritated or dehydrated skin.

Gardenia Jasminoides Meristem Cell Culture (Plant Stem Cell Extract, 0.5–2%)

Best For: Brightening, even skin tone, antioxidant protection.
What It Does: Provides gardenia stem cell compounds for gentle brightening and antioxidant defense.
Effectiveness: Improves radiance in 4–8 weeks with daily use.
Pair With: Vitamin C, Niacinamide, gentle exfoliants.

Avoid Using With: Strong Acids (<3.5 pH), harsh treatments.
Notes: Gentle; suits sensitive skin and K-beauty routines.

Glycyrrhiza Glabra (Licorice) Root Extract (Brightening Plant Extract, 0.5–5%)

Best For: Hyperpigmentation, dark spots, uneven tone.
What It Does: Inhibits tyrosinase with glabridin for brightening and anti-inflammatory benefits.
Effectiveness: Lightens in 4–8 weeks; optimal in 2–4 months with daily use.
Pair With: Vitamin C, Niacinamide, Sunscreen.
Avoid Using With: Strong Acids, harsh treatments.
Notes: Gentle alternative to hydroquinone; requires sun protection.

Hippophae Rhamnoides Callus Culture Extract (Sea Buckthorn) (Plant Stem Cell Extract, 0.5–2%)

Best For: Damaged skin, environmental stress, anti-inflammatory support.
What It Does: Delivers vitamins/antioxidants for repair and barrier protection.
Effectiveness: Repairs in 3–6 weeks; resilience in 2–3 months with daily use.
Pair With: Barrier Ingredients, Antioxidants, Moisturizers.
Avoid Using With: Strong Acids, harsh treatments.
Notes: Ideal for harsh climates; rich in omega fatty acids.

Leontopodium Alpinum Callus Culture Extract (Edelweiss) (Plant Stem Cell Extract, 0.5–2%)

Best For: Sensitive skin, UV protection, anti-inflammatory support.
What It Does: Provides UV-resistant compounds for protection and soothing.
Effectiveness: Soothes, protects in 2–4 weeks with daily use.
Pair With: Moisturizers, Antioxidants, Mineral Sunscreen.
Avoid Using With: Strong Actives, Exfoliants, Alcohol-Based Products.
Notes: Sustainable; ideal for high-altitude or sensitive skin.

Malus Domestica Fruit Cell Culture Extract (PhytoCellTec™) (Plant Stem Cell Extract, 0.5–2%)

Best For: Skin longevity, anti-aging prevention, cellular protection.
What It Does: Uses apple stem cells to protect skin cells and support renewal.
Effectiveness: Improves texture, vitality in 6–12 weeks with daily use.
Pair With: Antioxidants, Peptides, anti-aging ingredients.
Avoid Using With: Strong Oxidizing Agents, harsh treatments.
Notes: Focuses on long-term protection; sustainable Swiss apple source.

Rosa Damascena Callus Extracellular Vesicles (Plant Extracellular Vesicle Extract, 0.1–1%)

Best For: Sensitive skin, anti-aging, barrier repair, hydration.
What It Does: Delivers rose-derived vesicles with growth factors for regeneration and soothing.

Effectiveness: Soothes in 1–2 weeks; regenerates in 4–8 weeks with daily use.
Pair With: Hyaluronic Acid, Soothing Ingredients, sensitive skin products.
Avoid Using With: Strong Actives, harsh treatments.
Notes: Gentle; advanced vesicle delivery for sensitive skin.

Vitis Vinifera (Grape) Fruit Cell Extract (Plant Stem Cell Extract, 0.5–2%)

Best For: Antioxidant protection, UV damage prevention, anti-aging.
What It Does: Provides resveratrol/polyphenols for free radical defense and skin resilience.
Effectiveness: Enhances resilience in 4–6 weeks with daily use.
Pair With: Vitamin C, Sunscreen, Antioxidants.
Avoid Using With: Strong Oxidizing Agents, harsh treatments.
Notes: Strong preventative; synergistic with antioxidants.

Tips for Choosing Plant Stem Cell-Derived Actives

- **Anti-Aging Prevention**: Use Malus Domestica, Vitis Vinifera for antioxidant protection.
- Sensitive Skin: Choose Centella Asiatica, Leontopodium Alpinum, Rosa Damascena for soothing.
- **Brightening**: Opt for Glycyrrhiza Glabra, Gardenia Jasminoides for even tone.
- **Environmental Defense**: Prioritize Brassica Oleracea, Hippophae Rhamnoides for pollution/stress protection.
- **Hydration**: Select Cucumis Sativus for soothing moisture.
- **Sustainability**: All are lab-cultured, eco-friendly; pair with Sunscreen for protection.

PDRN and Polynucleotides

Polynucleotides (PN) and polydeoxyribonucleotides (PDRN) are regenerative actives derived from DNA fragments that are usually sourced from salmon or trout. Originally used in medicine for wound healing, tissue regeneration, and joint repair, they are now gaining attention in skincare for their ability to support deep skin renewal.

When applied to the skin, these DNA fragments break down into nucleotides and nucleosides, which are the essential building blocks your cells use to synthesize new DNA and RNA. But they don't just provide raw materials, they also activate the adenosine A2A receptor pathway, which helps:

- Reduce inflammation
- Stimulate angiogenesis (blood vessel formation)
- Promote collagen remodeling and tissue repair

In aesthetics, PDRN is best known in injectable form, where it's used to:

- Improve skin texture and elasticity
- Reduce scarring
- Accelerate post-procedure recovery

Topical forms, typically featuring PDRN or sodium DNA, are now appearing in advanced skincare products. However, because large DNA molecules have limited absorption through intact skin, pairing with professional treatments such as microneedling and microchanneling can significantly enhance penetration and effectiveness.

Unlike peptides or growth factors, polynucleotides work at the genetic level, signaling cells to repair, regenerate, and rebuild from the ground up. Their unique mechanism makes them one of the most exciting innovations in regenerative skincare, especially for those seeking skin recovery, scar remodeling, or improved resilience after professional procedures.

Adenosine (Nucleoside Anti-Aging Ingredient, 0.04–0.5%)

Best For: Fine lines, wrinkles, skin firmness, collagen stimulation.
What It Does: Enhances cellular energy, stimulates collagen, improves elasticity via repair pathways.
Effectiveness: Improves texture, firmness in 4–8 weeks with daily AM/PM use.
Pair With: Peptides, Niacinamide, Hyaluronic Acid.
Avoid Using With: Strong Acids, Alcohol-Based Products.
Notes: Well researched; supports natural repair; suitable for day/night use.

DNA (Deoxyribonucleic Acid) (Nucleic Acid Regenerative Ingredient, 0.1–2%)

Best For: Skin repair, hydration, anti-aging, wound healing.
What It Does: Provides nucleotides from salmon/plant sources for repair, forms hydrating films.
Effectiveness: Immediate hydration; regenerative benefits in 2–6 weeks with daily use.
Pair With: Hyaluronic Acid, Peptides, Antioxidants.
Avoid Using With: Strong Acids, degrading Enzymes, harsh treatments.
Notes: Salmon-derived; hydrates and supports cellular repair.

Hydrolyzed DNA (Nucleic Acid, 0.1–1%)

Best For: Hydration, anti-aging, skin conditioning.
What It Does: Moisturizes, supports skin barrier, and reduces the appearance of fine lines.
Effectiveness: Enhances skin hydration and smoothness within 4–6 weeks.
Pair With: Humectants (e.g., glycerin), peptides, niacinamide.
Avoid Using With: High-potency retinoids or exfoliating acids in the same step (if sensitivity occurs).
Notes: Broken down into smaller fragments for better absorption and enhanced moisturizing benefits.

Hydrolyzed RNA (Nucleic Acid, 0.1–1%)

Best For: Hydration, skin conditioning, anti-aging.
What It Does: Moisturizes, supports skin barrier function, and minimizes fine lines.
Effectiveness: Enhances hydration and skin smoothness within 4–6 weeks.
Pair With: Humectants (e.g., glycerin), niacinamide, peptides.
Avoid Using With: High-potency retinoids or exfoliating acids in the same step (if irritation occurs).
Notes: Processed into smaller nucleotide fragments for improved absorption and enhanced moisturizing benefits.

RNA (Nucleic Acid Cellular Support Ingredient, 0.1–1%)

Best For: Cellular metabolism, protein synthesis, anti-aging, hydration.
What It Does: Yeast/plant-derived RNA supports protein synthesis, hydrates skin.
Effectiveness: Metabolic benefits in 4–8 weeks with daily use; immediate hydration.
Pair With: Amino Acids, Peptides, hydrating ingredients.
Avoid Using With: RNA-degrading Enzymes, harsh treatments.
Notes: Yeast-derived; supports cellular function and hydration.

Sodium DNA (PDRN) (Nucleic Acid Hydrating Complex, 0.1–2%)

Best For: Intense hydration, skin plumping, anti-aging.
What It Does: Salmon/plant-derived DNA salt forms moisture-binding films, supports repair.
Effectiveness: Immediate plumping; regenerative support ongoing with daily use.
Pair With: Hyaluronic Acid, Glycerin, moisturizers.
Avoid Using With: Ingredients disrupting moisture binding, harsh treatments.
Notes: Enhanced stability; ideal for dry/dehydrated skin.
Sodium DNA (Various) (Nucleic Acid, 0.2–2%)
Best For: Wound healing, anti-aging, skin regeneration.
What It Does: Stimulates collagen production, promotes tissue repair, and enhances skin firmness.
Effectiveness: Visible improvement in skin texture and elasticity in 4–8 weeks.
Pair With: Growth factors, ceramides, hyaluronic acid.
Avoid Using With: Strong acids (e.g., glycolic acid) in the same step to avoid degradation.
Notes: Sourced sustainably from marine life (e.g., salmon milt); highly effective due to purified, depolymerized form (PDRN).

Sodium RNA (Nucleic Acid Cellular Support Complex, 0.1–1%)

Best For: Cellular function, anti-aging, hydration, metabolic support.
What It Does: Yeast/synthetic RNA salt enhances protein synthesis, provides hydration.
Effectiveness: Cellular benefits in 4–8 weeks; immediate hydration with daily use.
Pair With: Amino Acids, Peptides, hydrating complexes.
Avoid Using With: RNA-degrading Agents, harsh treatments.
Notes: Stable; supports cellular metabolism and hydration.

Tips for Choosing PDRN and Polynucleotides

- **Wound Healing:** Use Sodium DNA for rapid regeneration.
- **Hydration:** Choose Sodium DNA, DNA for intense moisture.
- **Anti-Aging:** Opt for Adenosine, Polynucleotide for firmness and repair.
- **Cellular Support:** Prioritize RNA, Sodium RNA for metabolism and protein synthesis.
- **Professional Use:** Pair with microneedling, microchanneling, or fractional skin rejuvenation treatments for enhanced delivery; use sterile products.
- **Storage:** Ensure proper handling to maintain stability, especially for PDRN/polynucleotides.

Complimentary & Supporting Ingredients

While active ingredients usually receive the spotlight in skincare discussions, the supporting cast of complementary ingredients plays an equally important role in determining whether a product is successful. These ingredients may not directly target specific skin concerns, but they create the foundation that makes effective skincare possible. They ensure products feel pleasant to use, remain stable over time, deliver actives efficiently, and provide sensory experiences that encourage consistent use.

Essential oils, with their complex molecular profiles and centuries of traditional use, provide a blend of regenerative, anti-inflammatory, and sensory benefits that enrich formulations. Their potency demands careful handling but rewards with natural therapeutic effects and elevated sensory appeal.

Synthetic fragrances provide long-lasting, versatile scent profiles that enhance product appeal, though they require mindful formulation due to potential sensitivities and allergenic risks.

Meanwhile, **sensory and aesthetic enhancers** such as texture modifiers, emollients, film-formers, and colorants shape how products feel, look, and perform, key factors in user satisfaction and routine adherence.

Finally, **pH adjusters** and **preservatives** work behind the scenes, ensuring product stability, safety, and optimal compatibility with the skin's natural environment. Understanding these complementary ingredients empowers smarter skincare choices and more effective formulation evaluation

The relationship between supporting and active ingredients is symbiotic rather than hierarchical. Preservatives protect the integrity of expensive actives. pH adjusters create an environment where acids can effectively exfoliate and where antioxidants can remain stable. Sensory modifiers ensure that potent treatments can be applied comfortably and evenly across the skin. Fragrance ingredients can influence mood and compliance, affecting how consistently products are used.

The most sophisticated skincare routine is only as effective as its weakest formulation link, and these supporting ingredients often determine where that weak point lies.

Essential Oils

Essential oils are nature's most concentrated botanical extracts, capturing the therapeutic essence of flowers, leaves, bark, roots, and seeds. Used for thousands of years in skin healing and regeneration, these volatile compounds are extracted through steam distillation or cold-pressing and remain a cornerstone of natural skincare.

What makes essential oils uniquely powerful is their molecular complexity and bioavailability. Each oil contains dozens to hundreds of active compounds that work synergistically to deliver potent effects ranging from anti-inflammatory and antimicrobial activity to cellular regeneration and barrier repair. Their small molecular size allows for deeper skin penetration, making them effective even in low concentrations.

However, essential oils must be used with precision and care. Their high potency means they can cause irritation or sensitization if applied undiluted. Many require dilution in carrier oils or incorporation into well-balanced formulas. Some, like citrus oils, can cause photosensitivity, while others may not be suitable for sensitive or reactive skin types. Safe and effective use starts with understanding each oil's properties, contraindications, and recommended concentrations.

There are many essential oils that can be used in natural healthcare and skincare products, so many that entire books have been written on the subject. The following are those you are most likely to find in skincare products.

Chamomilla Recutita (Matricaria) Flower Oil (Anti-Inflammatory Essential Oil, 0.1–1%)

Best For: Sensitive skin, irritation, redness, calming skin conditions.
What It Does: Provides chamazulene and bisabolol for anti-inflammatory and soothing effects, reduces redness.
Effectiveness: Calms irritation within 1–3 days with daily use; ongoing benefits with consistent application.
Pair With: Aloe Vera, Niacinamide, Hyaluronic Acid.
Avoid Using With: Undiluted use; strong Actives that may amplify sensitivity.
Notes: Gentle; ideal for sensitive skin; dilute with carrier oil (e.g., jojoba).

Citrus Aurantium Dulcis (Orange) Peel Oil (Antioxidant Essential Oil, 0.1–0.5%)

Best For: Brightening, oily skin, antioxidant protection.
What It Does: Contains limonene for antioxidant and mild exfoliating effects, balances oil production.
Effectiveness: Enhances radiance in 2–4 weeks with daily diluted use.
Pair With: Vitamin C, Niacinamide, lightweight moisturizers.
Avoid Using With: Sun exposure (photosensitizing); undiluted application.
Notes: Use at night; dilute to avoid irritation; refreshing scent.

Lavandula Angustifolia (Lavender) Oil (Calming Essential Oil, 0.1–1%)

Best For: Sensitive skin, acne, scalp health, stress-related skin issues.
What It Does: Provides linalool and linalyl acetate for anti-inflammatory, antimicrobial, and calming effects.
Effectiveness: Soothes skin/scalp in 1–3 days; improves acne in 2–4 weeks with daily use.
Pair With: Centella Asiatica, Hyaluronic Acid, Carrier Oils.
Avoid Using With: Undiluted use; high concentrations (risk of sensitization).
Notes: Versatile; calming aroma; dilute for safe use.

Melaleuca Alternifolia (Tea Tree) Leaf Oil (Antimicrobial Essential Oil, 0.1–1%)

Best For: Acne-prone skin, oily skin, scalp issues, blemish control.
What It Does: Provides terpinen-4-ol for strong antimicrobial and anti-inflammatory effects, targets acne-causing bacteria.
Effectiveness: Reduces blemishes in 1–2 weeks with daily diluted use.
Pair With: Salicylic Acid, Niacinamide, Aloe Vera.
Avoid Using With: Undiluted use; sensitive skin without patch-testing.
Notes: Potent; always dilute; strong scent may not suit all.

Mentha Piperita (Peppermint) Oil (Cooling Essential Oil, 0.1–0.5%)

Best For: Scalp health, irritation, cooling sensation, circulation support. Also used for lip plumping.
What It Does: Contains menthol for cooling, anti-inflammatory, and circulation-enhancing effects; soothes scalp itch.
Effectiveness: Immediate cooling; scalp benefits in 1–2 weeks with daily use.
Pair With: Rosemary Oil, Carrier Oils, Hydrating Ingredients.
Avoid Using With: Undiluted use; near eyes or mucous membranes.
Notes: Refreshing; dilute to avoid tingling or irritation.

Rosmarinus Officinalis (Rosemary) Leaf Oil (Antioxidant Essential Oil, 0.1–1%)

Best For: Scalp health, hair growth, oily skin, antioxidant support.
What It Does: Delivers rosmarinic acid for antioxidant, antimicrobial, and circulation-enhancing effects; supports hair growth.
Effectiveness: Improves scalp health, hair density in 4–8 weeks with daily use.
Pair With: Peppermint Oil, Biotin, Niacinamide.
Avoid Using With: Undiluted use; high concentrations (may irritate).
Notes: Stimulates scalp; natural preservative; dilute properly.

Tips for Choosing Essential Oils

- **Acne-Prone Skin**: Use Tea Tree Oil or Lavender Oil for antimicrobial effects.
- **Sensitive Skin**: Choose Chamomile Oil or Lavender Oil for gentle soothing.
- **Scalp Health**: Opt for Rosemary Oil or Peppermint Oil for circulation and growth.
- **Brightening**: Select Orange Peel Oil for radiance and oil control.

- **Dilution**: Always dilute with carrier oils (e.g., jojoba, argan) at 1–2% to avoid irritation.
- **Patch-Test**: Always Test on small area to check for sensitivity; avoid sun exposure with photosensitizing oils, especially citrus.

Synthetic Fragrances

Synthetic fragrances are lab-created aromatic compounds used extensively in skincare to add pleasant scent or mask undesirable odors. Unlike natural essential oils, synthetic fragrances are typically derived from petrochemicals and engineered to deliver longer-lasting, more intense, and highly customizable scent profiles.

While they provide versatility and stability in formulations, synthetic fragrances are also among the most common causes of skin irritation and allergic reactions, particularly for those with sensitive or compromised skin. One of the major concerns is the lack of ingredient transparency: brands can list complex mixtures of dozens, or even hundreds of aromatic compounds under a single umbrella term like "fragrance" or "parfum", making it difficult for consumers to identify potential allergens.

Some synthetic fragrance compounds also serve secondary functions in skincare, acting as preservatives, UV filters, or antimicrobial agents. However, their multi-functionality doesn't diminish their potential to cause contact dermatitis, redness, or flare-ups, particularly in individuals with eczema, rosacea, or fragrance allergies.

To address safety concerns, the European Union mandates disclosure of 26 specific fragrance allergens (such as limonene, linalool, and geraniol) when they exceed 0.001% in leave-on products or 0.01% in rinse-off formulas. This regulation supports more informed decision-making for sensitive users and reinforces the importance of label literacy in skincare.

Alpha-Isomethyl Ionone (Synthetic Violet Fragrance, 0.01–0.5%)

Best For: Luxury products, powdery violet scents, floral sophistication.
What It Does: Adds warm, floral, violet-like notes for a premium sensory experience.
Effectiveness: Long-lasting scent; requires EU allergen disclosure above thresholds.
Pair With: Floral Blends, gentle moisturizers, evening products.
Avoid Using With: Sensitive Skin, fragrance-free formulations.
Notes: Potential allergen; use sparingly in facial products.

Amyl Cinnamal (Synthetic Jasmine-Like Fragrance, 0.01–0.5%)

Best For: Sweet floral scents, feminine products, jasmine appeal.
What It Does: Provides jasmine-like floral notes, enhancing product desirability.
Effectiveness: Stable scent; EU allergen labeling required above limits.
Pair With: Floral Ingredients, gentle cleansers, moisturizers.
Avoid Using With: Sensitive Skin, fragrance-sensitive users.
Notes: Sweet and floral; patch-test for sensitive skin.

Benzyl Salicylate (UV-Absorbing Floral Fragrance, 0.01–0.5%)

Best For: Sunscreens, SPF products, mild floral scents.
What It Does: Provides sweet floral scent and UV absorption for sun protection.
Effectiveness: Moderate fragrance/UV protection; potential allergen.
Pair With: Sunscreen Actives, Moisturizers, floral blends.
Avoid Using With: Evening Products, highly sensitive skin.
Notes: Dual-purpose; use in daytime SPF formulations.

Citral (Bright Lemony Fragrance, 0.01–0.2%)

Best For: Fresh citrus scents, morning products, energizing formulations.
What It Does: Delivers vibrant lemony notes for uplifting sensory experience.
Effectiveness: Strong citrus scent; photosensitivity risk, requires EU labeling.
Pair With: Morning Cleansers, Citrus Blends, lightweight serums.
Avoid Using With: Evening Use, Sun Exposure, sensitive skin.
Notes: Refreshing; use with sunscreen to avoid photosensitivity.

Citronellol (Rose-Like Floral Fragrance, 0.01–0.5%)

Best For: Rose-scented products, feminine appeal, luxury formulations.
What It Does: Mimics rose fragrance, adding elegance to scent profiles.
Effectiveness: Stable, appealing floral scent; EU allergen disclosure needed.
Pair With: Floral Blends, Moisturizers, premium serums.
Avoid Using With: Sensitive Skin, fragrance-allergic users.
Notes: Sophisticated; test for sensitivity in facial use.

Coumarin (Sweet Vanilla-Like Fragrance, 0.01–0.5%)

Best For: Warm scents, evening products, gourmand profiles.
What It Does: Provides sweet, vanilla-like notes with hay-like nuances for comfort.
Effectiveness: Long-lasting scent; potential allergen, requires EU labeling.
Pair With: Evening Creams, Body Products, warm blends.
Avoid Using With: Sensitive Skin, fragrance-free products.
Notes: Comforting; use cautiously for sensitive users.

Geraniol (Sweet Rose-Like Fragrance, 0.01–0.5%)

Best For: Floral elegance, rose-scented products, feminine formulations.
What It Does: Delivers sweet, rose-like scent for luxurious sensory appeal.
Effectiveness: Stable, elegant fragrance; EU allergen labeling required.
Pair With: Floral Blends, Luxury Serums, feminine products.
Avoid Using With: Sensitive Skin, fragrance-sensitive users.
Notes: Broad appeal; patch-test for sensitive skin.

Hexyl Cinnamal (Jasmine-Like Floral Fragrance, 0.01–0.5%)

Best For: Soft jasmine scents, feminine products, evening routines.

What It Does: Adds elegant jasmine-like notes for appealing fragrance profiles.
Effectiveness: Stable, luxurious scent; potential allergen, requires EU labeling.
Pair With: Floral Blends, Evening Moisturizers, luxury products.
Avoid Using With: Sensitive Skin, fragrance-free formulations.
Notes: Sophisticated; use sparingly in sensitive areas.

Linalool (Floral Lavender-Like Fragrance, 0.01–0.5%)

Best For: Relaxing scents, evening products, broad-appeal formulations.
What It Does: Provides versatile lavender-like floral notes for relaxation.
Effectiveness: Stable, widely appealing; common allergen, requires EU labeling.
Pair With: Evening Products, Moisturizers, calming blends.
Avoid Using With: Highly Sensitive Skin, fragrance-allergic users.
Notes: Versatile; test for sensitivity despite good tolerance.

Parfum (Fragrance) (Synthetic Fragrance Blend, 0.1–1%)

Best For: Scent enhancement, odor masking, sensory appeal.
What It Does: Complex blend of aromatics for consistent, appealing scents.
Effectiveness: Enhances sensory experience; irritation varies by blend.
Pair With: Ceramides, Niacinamide, Panthenol, Antioxidants.
Avoid Using With: Strong Actives (AHAs, retinoids), Sensitive Skin.
Notes: Common allergen; use in low concentrations for safety.

Tips for Choosing Synthetic Fragrances

- **Luxury Appeal**: Use Alpha-Isomethyl Ionone, Citronellol, Geraniol for floral elegance.
- **Fresh Scents**: Choose Citral, Linalool for uplifting, clean notes.
- **Warm Profiles**: Opt for Coumarin, Hexyl Cinnamal for comforting scents.
- **Sensitive Skin**: Minimize use or select Parfum with soothing actives; patch-test always.
- **Regulatory** Compliance: Check EU allergen labeling for listed fragrances above thresholds.

Solubilizers and Emulsifying Agents

Solubilizers are the unsung heroes behind clear, stable skincare formulations. They dissolve oil-based ingredients, like essential oils, fragrances, or vitamin E into water-based products. Without solubilizers, these oils would separate and float on top, compromising both product stability and effectiveness. Solubilizers are essential in creating oil-free cleansers, micellar waters, and other water-based formulations that deliver oil-soluble actives evenly across the skin.

Emulsifying agents serve a similar but distinct function. They bind oil and water phases together to form smooth, uniform mixtures like creams, lotions, and cleansing milks. By reducing the surface tension between oil and water, emulsifiers allow these two naturally incompatible substances to blend and remain stable over time. Without them, many of the skincare textures we rely on daily would separate into unusable layers.

Together, solubilizers and emulsifiers make skincare both effective and enjoyable, ensuring consistent performance, elegant texture, and reliable results.

Caprylyl/Capryl Glucoside (Natural Non-Ionic Solubilizer, 1–10%)

Best For: Sensitive skin, natural formulations, children's products.
What It Does: Solubilizes essential oils and fragrances, derived from sugars and coconut for gentle clarity.
Effectiveness: Achieves clear solutions at 3:1–10:1 (carrier oils) or 4:1–6:1 (essential oils) ratios.
Pair With: Essential Oils, Natural Fragrances, Surfactants.
Avoid Using With: High pH (>10) formulations.
Notes: ECOCERT-approved; pre-blend for best clarity.

Cetearyl Alcohol (Co-Emulsifier/Thickener, 2–8%)

Best For: All skin types, creams, lotions, texture enhancement.
What It Does: Fatty alcohol that stabilizes emulsions, adds creamy texture, non-drying.
Effectiveness: Enhances viscosity, stability in 2–4 weeks with daily use formulations.
Pair With: Cetearyl Glucoside, Glycerin, Ceramides.
Avoid Using With: Insufficient primary emulsifiers (risk of separation).
Notes: Non-irritating; staple in premium creams.

Cetearyl Glucoside (Non-Ionic Sugar-Based Emulsifier, 2–5%)

Best For: Clean beauty, sensitive skin, luxury creams.
What It Does: Plant-derived emulsifier for stable, silky oil-in-water emulsions.
Effectiveness: Stable emulsions at 2–5%; optimal with cetearyl alcohol.
Pair With: Cetearyl Alcohol, Glycerin, Natural Oils.
Avoid Using With: Extreme pH (<3 or >10), high cationic polymers.
Notes: ECOCERT/COSMOS-approved; part of Montanov® 68.

Glyceryl Stearate (Emulsifying Agent, 1–10%)

Best For: Natural formulations, skin conditioning, clean beauty.
What It Does: Creates stable, non-greasy oil-in-water emulsions with conditioning benefits.
Effectiveness: Emulsifies up to 50% silicones at 1–10% for daily use.
Pair With: PEG-100 Stearate, Botanical Extracts, Vitamins.
Avoid Using With: Very low pH (<3) formulations.
Notes: ECOCERT-approved when non-ethoxylated; versatile.

Hydrogenated Lecithin (Phospholipid Emulsifier, 0.5–5%)

Best For: Cosmeceuticals, sensitive skin, active delivery systems.

What It Does: Forms biomimetic lamellar structures, enhances active penetration, supports barrier.
Effectiveness: Stable at 2–3% for lamellar systems; enhances delivery in weeks.
Pair With: Retinol, Peptides, Ceramides, Niacinamide.
Avoid Using With: High ionic surfactants, extreme pH.
Notes: Premium; stable for high-end formulations.

Polysorbate 20 (Solubilizer, 0.5–10%)

Best For: Micellar waters, makeup removers, clear formulations.
What It Does: Solubilizes oils in water-based systems, ideal for cleansing.
Effectiveness: Clear solutions at 0.5–5% (leave-on) or up to 10% (rinse-off).
Pair With: Botanical Oils, Hyaluronic Acid, Actives.
Avoid Using With: High oil loads without co-emulsifiers, clays.
Notes: Common in professional cleansers; gentle.

Polysorbate 80 (Solubilizer/Emulsifier, 0.5–10%)

Best For: Essential oil solubilization, stable microemulsions.
What It Does: Creates microemulsions, solubilizes fragrances and hydrophobic actives.
Effectiveness: Stable at 0.5–5% (leave-on) or up to 10% (rinse-off).
Pair With: Essential Oils, Fragrances, Fatty Alcohols.
Avoid Using With: High electrolytes (destabilizes emulsions).
Notes: Multifunctional; common in serums and cleansers.

Sucrose Stearate (Natural Sugar Ester Emulsifier, 1–5%)

Best For: K-beauty, sensitive skin, lightweight emulsions.
What It Does: Creates stable, light oil-in-water emulsions with conditioning.
Effectiveness: Stable, fresh texture at 1–5% for daily use.
Pair With: Lightweight Oils, Hyaluronic Acid, Botanical Extracts.
Avoid Using With: Very acidic pH (<3), high wax loads.
Notes: ECOCERT-approved; popular in Asian beauty.

Tips for Choosing Solubilizers and Emulsifying Agents

- **Natural Formulations**: Use Caprylyl/Capryl Glucoside, Cetearyl Glucoside, Sucrose Stearate for eco-friendly products.
- **Sensitive Skin**: Choose Hydrogenated Lecithin, Cetearyl Glucoside for gentle, biomimetic emulsions.
- **Clear Solutions**: Opt for Polysorbate 20, Polysorbate 80 for micellar waters or toners.
- **Luxury Textures**: Select Cetearyl Alcohol, Glyceryl Stearate for creamy, stable formulations.
- **Stability**: Pair with co-emulsifiers (e.g., Cetearyl Alcohol with Cetearyl Glucoside) for robust emulsions.
- **Patch-Test**: Always patch test for sensitivity in high-active or sensitive skin formulations.

Sensory Modifiers

Not every ingredient in skincare is designed to repair, protect, or transform the skin; some are included to influence how a product feels, smells, or behaves during use. Sensory modifiers are a class of ingredients that shape a product's texture, slip, absorption rate, fragrance, and after-feel. While they don't deliver direct biological benefits to skin cells, they greatly affect how a product is experienced, and, by extension, how likely someone is to use it consistently.

These ingredients range from lightweight silicones that create a silky finish, to polymers that add a cushiony texture, to natural oils, to waxes and butters that contribute richness and glide. Fragrance compounds and cooling agents like menthol also fall into this category, enhancing the sensory appeal of a formulation.

Though sometimes dismissed as "cosmetic extras," sensory modifiers play an important role in product success. A formula that feels greasy, sticky, or otherwise unpleasant is less likely to be used, no matter how potent its active ingredients may be.

That said, some sensory modifiers may cause issues for certain individuals. Fragrance compounds can trigger irritation in sensitive skin, and heavy silicones or waxes may contribute to buildup in acne-prone complexions. For these reasons, evaluating sensory ingredients is just as important as analyzing actives, especially for those with reactive or problem-prone skin.

Butylene Glycol (Humectant and Solvent, up to 10%)

Best For: Texture improvement, spreadability, absorption enhancement.
What It Does: Enhances product glide, hydrates, and aids active delivery with a non-greasy feel.
Effectiveness: Improves formulation stability, sensory feel immediately at low concentrations.
Pair With: Emollients, Actives, lightweight creams.
Avoid Using With: Excessive use on very sensitive skin (potential mild irritation).
Notes: Versatile; widely used for smooth application.

Cetearyl Ethylhexanoate (Lightweight Emollient Ester, 2–10%)

Best For: Silky feel, non-greasy formulations, spreadability.
What It Does: Provides smooth, lightweight emolliency with a non-greasy finish.
Effectiveness: Enhances skin softness, spreadability immediately at 2–10%.
Pair With: Silicones, Humectants, lightweight serums.
Avoid Using With: Emulsion-destabilizing ingredients.
Notes: Ideal for elegant, non-heavy textures.

Cyclopentasiloxane (Volatile Silicone, 1–10%)

Best For: Quick-drying products, luxurious slip, temporary smoothing.

What It Does: Provides silky application, evaporates for a dry, smooth finish.
Effectiveness: Immediate smoothing enhances active delivery; effects last minutes.
Pair With: Oil-Soluble Actives, Heavy Ingredients, sunscreens.
Avoid Using With: Thin water-based gels, very oily skin.
Notes: Volatile; ideal for lightweight formulations.

Dimethicone (Silicone Barrier Former, 1–10%)

Best For: Skin smoothing, water loss prevention, line filling.
What It Does: Forms a breathable barrier, smooths skin, reduces water loss.
Effectiveness: Immediate texture improvement reduces TEWL up to 8 hours.
Pair With: Glycerin, Hyaluronic Acid, Antioxidants.
Avoid Using With: Oil-based ingredients, strong emulsifiers.
Notes: Non-comedogenic; widely used in moisturizers.

Hydroxyethylcellulose (Plant-Derived Thickener, 0.5–2%)

Best For: Texture modification, gel stability, water-based products.
What It Does: Thickens water-based formulations, creates stable gels or creams.
Effectiveness: Stable texture across pH/temperature, effective immediately.
Pair With: Water-Soluble Actives, Glycerin, Botanical Extracts.
Avoid Using With: High alcohol, strongly ionic ingredients.
Notes: Natural; versatile for gels and serums.

Menthyl Lactate (Gentle Cooling Agent, 0.5–1%)

Best For: Subtle cooling, sensitive skin, refreshing products.
What It Does: Provides gentle, long-lasting cooling without irritation.
Effectiveness: Cooling for 30–60 minutes; minimal sensitivity risk.
Pair With: Aloe Vera, Centella Asiatica, hydrating gels.
Avoid Using With: Strong acids, retinoids, occlusive ingredients.
Notes: Gentle alternative to menthol; ideal for sensitive skin.

Neopentyl Glycol Diheptanoate & Heptanoate Esters (Lightweight Ester Emollients, 3–20%)

Best For: Oily/combination skin, dry-touch products, non-comedogenic.
What It Does: Provides light, silky emolliency with powdery finish.
Effectiveness: Immediate luxurious feel, low comedogenicity at 3–20%.
Pair With: Hyaluronic Acid, Niacinamide, Squalane.
Avoid Using With: None; highly stable.
Notes: Popular in oil-free moisturizers, primers.

Silicone Quaternium-16/Glycidoxy Dimethicone Crosspolymer (Silicone Polymer, 0.5–5%)

Best For: Skin conditioning, texture enhancement, moisture retention.
What It Does: Forms a breathable, protective film on skin, smooths texture, and locks in hydration.

Effectiveness: Improves skin feel and softness immediately; long-term hydration with consistent use.
Pair With: Humectants (e.g., hyaluronic acid), emollients, lightweight serums.
Avoid Using With: Heavy occlusives in the same step (may reduce breathability).
Notes: A silicone-based polymer that enhances product spreadability; non-comedogenic in most formulations, ideal for dry or sensitive skin.

Tips for Choosing Sensory Modifiers

- **Luxury Textures**: Use Neopentyl Glycol Diheptanoate for silky, premium feel.
- **Lightweight Formulations**: Choose Cyclopentasiloxane, Cetearyl Ethylhexanoate for non-greasy finishes.
- **Hydration Support**: Pair Butylene Glycol, Dimethicone with humectants for moisture retention.
- **Sensitive Skin**: Opt for Menthyl Lactate, Hydroxyethylcellulose for gentle sensory effects.
- **Stability**: Ensure compatibility with formulation pH and emulsifiers for consistent texture.
- **Patch-Test**: Always patch test for sensitivity, especially with cooling agents or silicones in reactive skin.

Colorants and Tints

Colorants and tints are ingredients added to skincare and cosmetic products to change or enhance visual appearance, either of the product itself or the skin. While they don't repair or rejuvenate the skin, they play a powerful role in how a product is perceived and used. A cream may look more appealing with a soft hue, a serum might seem more luxurious, or a tinted moisturizer may provide instant tone correction or visual improvement.

These pigments range from mineral-based options like iron oxides and titanium dioxide, to synthetic dyes and lakes, to natural colorants such as beetroot extract or chlorophyll. In tinted formulations, especially sunscreens and moisturizers, colorants often serve a dual purpose: improving cosmetic appearance and enhancing performance. For example, iron oxides not only provide a natural-looking tint but also help protect against visible light, which can trigger hyperpigmentation, particularly in melanin-rich or melasma-prone skin.

The CI (Color Index) number for each pigment ingredient is a unique, standardized ID so you can tell exactly which colorant is present, regardless of brand names or marketing language. A color's unique CI Number makes it easier to

- Verify shade families and performance cues. For example, iron oxides CI 77491/77492/77499 blend to create skin-true tints; titanium dioxide is CI 77891.

- Verify the colorant for region-specific approvals and use restrictions, something especially for eye and lip products.

- Avoid allergens or animal-derived pigments, for example carmine CI 75470 a natural red dye derived from cochineal insects.

- Spot functional benefits like visible-light protection from iron oxides. For example, Chlorophyllin–Copper Complex (CI 75810) identifies a specific green plant-derived colorant, so you can confirm the hue source and compare similar formulas with confidence.

Also note that global labels often include a "±/May contain" list of CI numbers to show shade variations across a product line.

Beta-Carotene (CI 40800) (Antioxidant Colorant, 0.01–0.5%)

Best For: Antioxidant protection, natural yellow-orange coloring, UV stress defense.
What It Does: Provides yellow-orange pigmentation and antioxidant benefits, neutralizing free radicals.
Effectiveness: Immediate antioxidant action; visible skin protection in 2–4 weeks with daily use.
Pair With: Vitamin C, Sunscreen, Antioxidants, natural formulations.
Avoid Using With: Strong Oxidizing Agents, harsh treatments.
Notes: Natural; may impart slight orange tint on light skin.

Chlorophyllin-Copper Complex (CI 75810) (Antioxidant Colorant, 0.01–0.1%)

Best For: Acne-prone skin, green coloring, anti-inflammatory support.
What It Does: Provides green pigmentation, antioxidant, and antimicrobial benefits for acne and inflammation.
Effectiveness: Soothes in 1–2 weeks; ongoing acne benefits with daily use.
Pair With: Salicylic Acid, Niacinamide, natural acne treatments.
Avoid Using With: Strong Acids, irritating treatments.
Notes: Semi-synthetic; ideal for natural, acne-focused products.

Iron Oxides (CI 77491, CI 77492, CI 77499) (Mineral Colorant, 1–15%)

Best For: Tinted skincare, sensitive skin, mild UV protection.
What It Does: Provides red, yellow, black pigmentation and gentle UV protection.
Effectiveness: Immediate color coverage; mild UV protection with daily use.
Pair With: Titanium Dioxide, Zinc Oxide, Moisturizers, tinted sunscreens.
Avoid Using With: None; highly compatible.
Notes: Safe for sensitive skin; enhances mineral makeup.

Titanium Dioxide (CI 77891) (Mineral UV Filter and Colorant, 5–25%)

Best For: Sun protection, sensitive skin, white/light coloring.
What It Does: Reflects UV rays, provides white pigmentation, stable for sensitive skin.
Effectiveness: Immediate broad-spectrum UV protection; no wait time needed.
Pair With: Zinc Oxide, Iron Oxides, Antioxidants, Silicones.
Avoid Using With: Unstabilized Avobenzone, surface charge-altering ingredients.

Notes: May cause white cast; nano/non-nano options available.

Zinc Oxide (CI 77947) (Mineral UV Filter and Anti-Inflammatory, 2–25%)

Best For: Broad-spectrum protection, sensitive/acne-prone skin, soothing.
What It Does: Provides white pigmentation, broad-spectrum UV protection, anti-inflammatory benefits.
Effectiveness: Immediate UV protection, soothes in days, improves resilience in 1–2 weeks.
Pair With: Titanium Dioxide, Niacinamide, Copper Peptides, Panthenol.
Avoid Using With: Ascorbic Acid, high AHAs.
Notes: Gentle; antimicrobial; minimizes white cast in modern formulations.

Tips for Choosing Colorants and Tints

- **Natural Coloring**: Use Beta-Carotene, Chlorophyllin-Copper Complex for eco-friendly, functional pigmentation.
- **Sensitive Skin**: Choose Iron Oxides, Titanium Dioxide, Zinc Oxide for safe, non-reactive coverage.
- **Sun Protection**: Opt for Titanium Dioxide, Zinc Oxide for broad-spectrum UV defense.
- **Acne-Prone Skin**: Select Chlorophyllin-Copper Complex, Zinc Oxide for anti-inflammatory, antimicrobial benefits.
- **Formulation**: Blend Iron Oxides with Titanium Dioxide/Zinc Oxide for natural-looking tinted sunscreens.
- **Safety**: Always patch-test for sensitive skin; ensure compatibility with actives to maintain stability.

---

pH Adjusters

Every skincare formula has an optimal pH range where its ingredients are most stable, effective, and least likely to irritate the skin. Human skin naturally maintains a slightly acidic pH, typically between 4.5 and 5.5, which supports a healthy barrier and microbiome. Skincare products formulated close to this range are generally more compatible with the skin and less likely to cause disruption.

pH adjusters are ingredients used to fine-tune a product's acidity or alkalinity, ensuring the formula remains stable, effective, and comfortable to use. While they don't deliver direct skin benefits, they're essential for preserving the performance of key actives like AHAs, vitamin C, peptides, and preservatives. A formula that's too acidic can cause irritation, while one that's too alkaline can degrade sensitive ingredients and compromise the skin barrier.

Common pH-lowering agents include citric acid and lactic acid, while pH-raising agents include sodium hydroxide and tromethamine (triethanolamine). These adjusters are typically used in very small concentrations and often appear at the end of ingredient lists, not because they're unimportant, but because only small amounts are needed to bring the formula into the desired pH range.

By keeping the formulation within an ideal pH window, these ingredients work behind the scenes to help ensure skincare product safety, stability, and effectiveness.

Citric Acid (pH Adjuster and Antioxidant, 0.01–1%)

Best For: pH reduction, antioxidant support, product stabilization.
What It Does: Lowers pH, chelates metal ions, provides mild antioxidant benefits.
Effectiveness: Immediate pH adjustment; ongoing stability/antioxidant effects.
Pair With: Vitamin C, Antioxidant Serums, Chelators.
Avoid Using With: Highly alkaline ingredients, metal-dependent actives.
Notes: Natural, GRAS; essential for vitamin C serum stability.

Magnesium Hydroxide (pH Adjuster and Anti-Inflammatory, 0.5–5%)

Best For: Alkaline adjustment, oil control, deodorants, soothing.
What It Does: Gently raises pH, absorbs oil, reduces inflammation.
Effectiveness: Gradual pH adjustment; ongoing soothing/oil control.
Pair With: Oil-Control Actives, Anti-Inflammatory Ingredients.
Avoid Using With: Strong acids, precise acidic pH needs.
Notes: Natural; dual-purpose for clean beauty formulations.

Sodium Bicarbonate (pH Adjuster and Gentle Abrasive, 0.5–3%)

Best For: Alkaline adjustment, gentle exfoliation, deodorizing.
What It Does: Raises pH, provides mild exfoliation, neutralizes odors.
Effectiveness: Gradual pH adjustment; immediate exfoliation/deodorizing.
Pair With: Gentle Cleansers, Exfoliating Scrubs, deodorants.
Avoid Using With: Strong acids, very sensitive skin.
Notes: Food-grade; natural, dual-purpose for clean formulations.

Sodium Citrate (pH Adjuster and Chelating Agent, 0.1–1%)

Best For: pH buffering, stabilization, sensitive skin products.
What It Does: Buffers pH, chelates metal ions for formulation stability.
Effectiveness: Consistent pH buffering; ongoing chelation benefits.
Pair With: Vitamin C, Antioxidants, Sensitive Skin Actives.
Avoid Using With: Metal-dependent actives.
Notes: Natural; key for antioxidant formulation stability.

Tartaric Acid (pH Adjuster and Antioxidant, 0.1–1%)

Best For: Acidic pH adjustment, antioxidant support, natural formulations.
What It Does: Lowers pH, provides mild antioxidant benefits.
Effectiveness: Immediate pH adjustment; ongoing antioxidant protection.
Pair With: Grape-Derived Actives, Antioxidants, gentle acids.
Avoid Using With: Highly alkaline ingredients.

Notes: Natural, grape-derived; adds clean beauty appeal.

Triethanolamine (pH Adjuster and Emulsifying Agent, 0.1–2%)

Best For: pH adjustment, emulsification, acidic ingredient neutralization.
What It Does: Neutralizes acids, stabilizes emulsions, adjusts pH.
Effectiveness: Rapid pH adjustment; ongoing emulsion stability.
Pair With: Acidic Actives, Carbomer Gels, emulsions.
Avoid Using With: Amine-sensitive formulations, nitrosamine risks.
Notes: Multifunctional; requires careful use to avoid nitrosamines.

Tromethamine (pH Adjuster and Buffer, 0.1–1%)

Best For: Gentle pH adjustment, sensitive skin, stable buffering.
What It Does: Gently raises pH, maintains stable buffering.
Effectiveness: Gradual, controlled pH adjustment; stable buffering.
Pair With: Sensitive Skin Treatments, Gentle Acids, buffers.
Avoid Using With: Strong acids overwhelming buffering capacity.
Notes: Gentle; ideal for sensitive skin formulations.

Trisodium Ethylenediamine Disuccinate (Chelating Agent and Stabilizer, 0.01–0.1%)

Best For: Metal ion chelation, preservative enhancement, stability.
What It Does: Binds metals, enhances preservative efficacy, stabilizes formulas.
Effectiveness: Immediate chelation; ongoing stability protection.
Pair With: Preservatives, Vitamin C, Antioxidant Serums.
Avoid Using With: Metal-dependent actives.
Notes: Biodegradable; eco-friendly alternative to EDTA.

Tips for Choosing pH Adjusters

- **Sensitive Skin**: Use Tromethamine, Sodium Citrate for gentle buffering.
- **Natural Formulations**: Opt for Citric Acid, Tartaric Acid, Magnesium Hydroxide for clean beauty appeal.
- **Antioxidant Stability**: Pair Sodium Citrate, Trisodium Ethylenediamine Disuccinate with vitamin C serums.
- **Deodorizing/Exfoliation**: Select Sodium Bicarbonate for dual-purpose effects.
- **Stability**: Ensure precise measurement for potent adjusters (Triethanolamine) to avoid over-neutralization.
- **Safety**: Patch-test formulations with strong bases; monitor for nitrosamine risks with amines.

Preservatives

Preservatives are among the most misunderstood and most essential ingredients in skincare. Their primary role is to prevent the growth of bacteria, mold, and yeast in formulations that contain water, oils, and nutrient-rich ingredients that are ideal environments for microbial growth. Without preservatives, even the most well-formulated product could become unsafe within days, risking infection and irritation rather than delivering benefits.

Preservatives also protect the integrity of active ingredients. Peptides, vitamins, and botanical extracts can degrade when exposed to microbes or oxidation, reducing both their safety and effectiveness. A well-designed preservative system keeps formulas stable, consistent, and effective from the first pump to the last.

Most skincare products rely on a combination of preservatives; no single compound is broad-spectrum enough to cover bacteria, fungi, and mold on their own. Common preservatives include phenoxyethanol, ethylhexylglycerin, potassium sorbate, and sodium benzoate. Preservative systems typically fall into these categories:

- **Synthetic Preservatives** Highly effective and widely used, examples include phenoxyethanol, chlorphenesin, and diazolidinyl urea. These provide strong, broad-spectrum antimicrobial protection and long shelf life.

- **Organic Acids & Their Salts** Gentler, more skin-friendly options like sorbic acid, benzoic acid, sodium benzoate, and potassium sorbate are effective against fungi and mold and commonly used in clean or natural formulations.

- **Multifunctional Boosters** Ingredients such as ethylhexylglycerin, caprylyl glycol, and hexanediol provide antimicrobial support while also acting as humectants, emollients, or surfactants.

- **Antioxidants** Compounds like tocopherol (vitamin E) and ascorbyl palmitate are not true preservatives, but they prevent oils from going rancid and extend product freshness.

Some preservatives, particularly parabens and formaldehyde-releasing agents, have raised health concerns. Parabens were once industry staples due to their effectiveness and low cost. However, studies suggesting potential hormone-disrupting activity led to consumer backlash and increased scrutiny, even though they remain approved by major regulatory bodies like the FDA, EU, and Japan. Today's formulators often favor organic acids and multifunctional ingredients to meet consumer demand for gentler alternatives while still maintaining safety and efficacy.

Any product containing water must include a preservative, this is not optional. Even at low, regulated levels, preservatives can sometimes cause irritation in very sensitive skin, so selecting mild, well-tolerated systems is essential. Preservatives may not be glamorous, but they are non-negotiable for safe, stable, and effective skincare.

Benzoic Acid (Antimicrobial Preservative, 0.1–0.5%)

Best For: Acidic formulations, natural preservation, microbial protection.
What It Does: Inhibits bacteria, yeasts, molds in low pH (<4.5) environments.
Effectiveness: Immediate antimicrobial action; sustained in acidic conditions.

Pair With: Sodium Benzoate, Acidic Formulations, Natural Preservatives.
Avoid Using With: High pH (>4.5), sensitive skin at high concentrations.
Notes: Natural; pH-dependent; may irritate sensitive skin.

Benzyl Alcohol (Antimicrobial Preservative and Solvent, 0.5–1%)

Best For: Gentle preservation, fragrance dissolution, natural systems.
What It Does: Provides antimicrobial protection, acts as solvent for fragrances.
Effectiveness: Immediate, sustained antimicrobial action; versatile solvent.
Pair With: Natural Preservatives, Fragrances, Essential Oils.
Avoid Using With: Alcohol-sensitive skin, very sensitive formulations.
Notes: Natural occurrence; dual-purpose; may irritate sensitive skin.

Caprylyl Glycol (Preservative Booster and Skin Conditioner, 0.1–1%)

Best For: Gentle preservation, skin conditioning, natural formulations.
What It Does: Boosts preservatives, provides emollient properties.
Effectiveness: Immediate preservative enhancement; ongoing conditioning.
Pair With: Preservatives, Natural Formulations, Humectants.
Avoid Using With: Glycol-sensitive formulations.
Notes: Multifunctional; ideal for clean beauty, sensitive skin.

Ethylhexylglycerin (Preservative Booster and Skin Conditioner, 0.1–1%)

Best For: Preservative enhancement, deodorizing, skin conditioning.
What It Does: Enhances preservatives, conditions skin, controls odor.
Effectiveness: Immediate enhancement; sustained conditioning benefits.
Pair With: Phenoxyethanol, Deodorizing Actives, Emollients.
Avoid Using With: Glycol-free formulations, glycerin-sensitive skin.
Notes: Synergistic with phenoxyethanol; excellent safety profile.

Phenoxyethanol (Antimicrobial Preservative, 0.5–1%)

Best For: Broad-spectrum preservation, reliable cosmetic protection.
What It Does: Protects against bacteria, yeasts, molds; highly compatible.
Effectiveness: Immediate, sustained antimicrobial action.
Pair With: Ethylhexylglycerin, Caprylyl Glycol, broad-spectrum systems.
Avoid Using With: Glycol ether-sensitive skin, glycol-free formulations.
Notes: Widely used; safe but may irritate sensitive skin.

Potassium Sorbate (Antimicrobial Preservative, 0.1–0.2%)

Best For: Yeast/mold prevention, natural preservation, acidic systems.
What It Does: Inhibits yeasts, molds; extends shelf life in low pH.
Effectiveness: Immediate, sustained in acidic conditions (<5.5).
Pair With: Sodium Benzoate, Acidic Formulations, Natural Systems.
Avoid Using With: High pH, conflicting pH requirements.
Notes: GRAS; natural; effective in acidic environments.

Sodium Benzoate (Antimicrobial Preservative, 0.1–0.5%)

Best For: Acidic formulations, natural preservation, bacterial/fungal control.
What It Does: Inhibits bacteria, yeasts, molds in low pH (<4.5).
Effectiveness: Immediate, sustained in acidic conditions.
Pair With: Potassium Sorbate, Vitamin C, Acidic Serums.
Avoid Using With: High pH, metal ions with vitamin C.
Notes: GRAS; natural; pH-dependent efficacy.

Tetrasodium Glutamate Diacetate (Chelating Agent and Stabilizer, 0.01–0.1%)

Best For: Metal ion chelation, preservative enhancement, eco-friendly stability.
What It Does: Binds metals, boosts preservatives, stabilizes formulations.
Effectiveness: Immediate chelation; ongoing stability protection.
Pair With: Preservatives, Vitamin C, Antioxidant Serums.
Avoid Using With: Metal-dependent actives.
Notes: Biodegradable; eco-friendly alternative to EDTA.

Tips for Choosing Preservatives

- **Natural Formulations**: Use Benzoic Acid, Potassium Sorbate, Sodium Benzoate for clean beauty appeal.
- **Sensitive Skin**: Opt for Caprylyl Glycol, Ethylhexylglycerin for gentle, conditioning preservation.
- **Broad-Spectrum Protection**: Pair Phenoxyethanol with Ethylhexylglycerin for robust coverage.
- **Acidic Systems**: Select Benzoic Acid, Sodium Benzoate, Potassium Sorbate for low pH efficacy.
- **Stability**: Use Tetrasodium Glutamate Diacetate to enhance preservative performance and formulation stability.
- **Safety**: Always patch-test for sensitive skin; avoid high concentrations in glycol-sensitive formulations.

GLOSSARY

Acid Mantle: The skin's mildly acidic surface film (pH approximately 4.7–5.5) that supports the microbiome and barrier enzymes. Harsh cleansers can disrupt it.

Active Ingredient (Cosmetic): An ingredient included primarily for a beneficial cosmetic effect (e.g., improving tone or supporting barrier function), distinct from regulated drug actives.

Active Ingredient (OTC Drug): A regulated active ingredient that must follow an FDA monograph and appear in a Drug Facts panel, such as sunscreen filters or acne treatments.

AHA (Alpha Hydroxy Acid): Water-soluble chemical exfoliants that promote desquamation by loosening bonds between dead skin cells. Lower pH increases potency. Examples include glycolic and lactic acid.

Allergen: A substance that can cause an allergic reaction, often manifesting as contact dermatitis, itching, or redness. Common cosmetic allergens include certain fragrances and preservatives.

Anhydrous: A water-free formulation (such as balms, oils, or sticks) that relies on oils, waxes, or silicones. Often more stable for water-sensitive active ingredients.

Antioxidant: A substance that neutralizes free radicals and helps prevent oxidation to protect formulas and reduce oxidative stress on skin. Common examples include vitamin C, vitamin E, and resveratrol.

Astringent: Ingredients that tighten skin and reduce oil, often used in toners.

Barrier Function (Skin Barrier): The skin's outermost protective layer (stratum corneum) composed of skin cells and lipids in a "brick and mortar" structure. It prevents water loss and blocks irritants, allergens, and pathogens from entering the skin.

BHA (Beta Hydroxy Acid): An oil-soluble chemical exfoliant that can penetrate pores, making it effective for treating blackheads and acne. Salicylic acid is the most common BHA.

Bioavailability: The fraction of an active ingredient that reaches its site of action in the skin, affected by vehicle, molecular size, solubility, and encapsulation.

Blue Light (HEV): High-energy visible light (approximately 400–500 nm) that may contribute to pigmentation issues in some skin types. Iron oxides can help attenuate it.

Botanical Extract: A plant-derived mixture containing multiple compounds, often used for antioxidant, anti-inflammatory, or soothing properties. Composition varies by extraction method and quality.

Broad-Spectrum (Sunscreen): A label indicating protection against both UVA and UVB radiation per regional test standards.

Carrier Oil: A base oil used to dilute more potent ingredients or deliver them into the skin. Examples include jojoba oil, squalane, and rosehip oil. These oils also provide their own emollient and nourishing benefits.

Chelating Agent (Chelator): Compounds that bind metal ions to stabilize formulations, improve preservative efficacy, and reduce catalyzed oxidation.

CI Number: (Color Index): An international classification system for colorants that assigns standardized pigment identification numbers. For example, CI 75810 identifies

Cleanser: A product designed to remove dirt, oil, makeup, and impurities from the skin.

Colorant/Tint: An ingredient added to change or enhance the visual appearance of a product or skin. Ranges from mineral pigments like iron oxides to synthetic dyes to natural extracts like beetroot. Some colorants, particularly iron oxides, also provide visible light protection.

Comedogenic: Describes the tendency of an ingredient or product to clog pores and potentially causes breakouts. Comedogenicity ratings (0-5 scale) indicate likelihood, though individual reactions vary.

Concentration: The percentage or amount of an ingredient in a formula. Higher concentrations aren't always better and can increase irritation risk. Effective concentrations vary by ingredient.

Corneocyte: A dead, flattened cell in the outermost layer of skin. Shedding is regulated by enzymes and pH.

Cosmeceutical: A marketing term (not a legal category) for cosmetics claiming stronger benefits. Still regulated as cosmetics in many regions.

Delivery System: The technology or method used to transport active ingredients into the skin. Examples include liposomes, encapsulation, and time-release polymers. Advanced delivery systems can improve stability, penetration, and efficacy.

Dermatitis: A general term for skin inflammation, often characterized by a red, itchy rash. Contact dermatitis is inflammation caused by direct contact with a substance (an irritant or an allergen).

Dermatologically Tested: Indicates a product has been tested on human skin, though standards vary and it doesn't guarantee safety for all users.

Drug Claim: A statement that implies the product treats or alters skin function, regulated by the FDA.

Drug Facts Panel: The standardized label required for U.S. OTC drugs listing active ingredients, purpose, uses, directions, and warnings.

Efficacy: The ability of a product or ingredient to produce the desired or intended result, especially in a measurable way.

Emollient: An ingredient that softens and smooths the skin's surface by filling in gaps between skin cells. Common emollients include plant oils, butters, silicones, and fatty alcohols like cetyl and stearyl alcohol.

Emulsifier: A substance that allows oil and water to mix and stay blended in a stable formula. Without emulsifiers, creams and lotions would separate into layers. Examples include lecithin, polysorbates, and cetearyl alcohol.

Emulsion: A stable mixture of two liquids that normally won't mix (like oil and water), achieved with the help of an emulsifier. Most creams, lotions, and milk cleansers are emulsions. Common types are O/W (oil-in-water, lighter feel) and W/O (water-in-oil, more occlusive).

Encapsulation: A delivery strategy using liposomes or polymer capsules that protects active ingredients and modulates release or reduces irritation.

Essential Oil: Concentrated plant oils used for fragrance or therapeutic effects but may cause irritation in sensitive skin.

Ester (Emollient Ester): A lightweight, fast-spreading emollient formed from an acid plus alcohol. Some may be comedogenic for certain individuals.

Exfoliant: An ingredient that removes dead skin cells from the surface. Physical exfoliants use particles or tools; chemical exfoliants use acids (AHAs, BHAs, PHAs) or enzymes to dissolve cellular bonds.

Fatty Acid: Lipid building blocks that are important for barrier lipids and emulsion structure. Essential fatty acids nourish and support the skin barrier.

Fatty Alcohol: Waxy alcohols (such as cetyl, stearyl, cetearyl) used as thickening emollients. Not the same as drying simple alcohols.

Formulation: The complete recipe of a skincare product, including all ingredients and their concentrations. Good formulation considers ingredient compatibility, stability, pH, texture, and efficacy.

Fragrance/Parfum: A blend of aromatic compounds used to scent products. Can be synthetic or natural and may trigger sensitivity. A common source of contact dermatitis for sensitive skin.

Free Radicals (ROS): Unstable molecules with unpaired electrons that damage skin cells through oxidation. Generated by UV exposure, pollution, and normal metabolism. Antioxidants neutralize free radicals.

Humectant: An ingredient that attracts and binds water to the skin. Common humectants include hyaluronic acid, glycerin, urea, and propanediol. They increase skin hydration when used with occlusives.

Hydrophilic: Water-loving. Describes ingredients that dissolve in or attract water.

Hyperpigmentation: Darkening of skin due to inflammation, UV/visible light exposure, or hormones. Types include post-inflammatory hyperpigmentation (PIH) and melasma. Managed with sunscreen, iron oxides, and pigment-modulating ingredients.

Hypoallergenic: A marketing term implying lower allergen risk. Not a guarantee of non-irritation and not strictly regulated.

INCI Name (International Nomenclature of Cosmetic Ingredients): The standardized naming system used on ingredient lists worldwide. For example, the INCI name for vitamin C is "Ascorbic Acid."

Irritant: A substance that causes a direct, non-allergic inflammatory reaction upon contact, often resulting in stinging, burning, redness, or discomfort.

Keratolytic: An ingredient that helps break down keratin to soften and shed dead skin, improving texture.

Lipid: A general term for fats and fat-soluble substances. In skincare, essential lipids like ceramides, cholesterol, and fatty acids are vital components of the skin barrier.

Lipophilic: Oil-loving. Describes ingredients that dissolve in or are attracted to oils and fats.

Liposome: A phospholipid vesicle used to encapsulate and deliver active ingredients with improved stability and penetration.

Melanin: The pigment produced by melanocytes. Overproduction leads to hyperpigmentation.

Microbiome (Skin Microbiota): The community of microorganisms on skin that interact with the barrier and immune system. Supported by mild pH and gentle cleansing.

Micronized/Nano: Particle size reduction to improve dispersion and feel. Nanomaterials may have specific regulatory labeling requirements.

Mineral (Inorganic) Sunscreen: UV filters zinc oxide and titanium dioxide that reflect, scatter, and absorb UV radiation. Often micronized to reduce white cast.

Moisturizer: A product designed to hydrate and protect the skin using humectants (to attract water), emollients (to smooth skin), and occlusives (to seal moisture in).

Monograph (OTC): The FDA's rulebook for certain OTC drug actives that specifies allowable concentrations, indications, and labeling.

Natural Ingredient: Derived from plant, mineral, or animal sources. Not synonymous with safe or gentle.

Natural Moisturizing Factor (NMF): Endogenous humectants (amino acids, PCA, urea) within skin cells that maintain hydration.

Non-Comedogenic: Indicates ingredients or products formulated to avoid clogging pores. Not a regulated claim and may vary by individual.

Occlusive: An ingredient that forms a protective barrier on the skin's surface to prevent water loss. Examples include petrolatum, mineral oil, beeswax, and dimethicone. Most effective when applied over humectants.

Oil-Free: Indicates a product does not contain oils, often preferred by acne-prone or oily skin types.

Oxidative Stress: Cellular damage caused by an imbalance between free radicals and the body's ability to neutralize them with antioxidants. Contributes to photoaging and inflammation.

Penetration Enhancer: An ingredient that helps other ingredients penetrate deeper into the skin. Examples include propanediol, certain alcohols, and dimethyl isosorbide.

pH (Potential of Hydrogen): A measure of acidity or alkalinity on a scale of 0-14. Skin's natural pH is around 4.5-5.5 (slightly acidic). Product pH affects ingredient stability, efficacy, and skin compatibility.

Preservative: An ingredient that prevents microbial growth and extends product's shelf life. Common preservatives include phenoxyethanol, parabens, and benzyl alcohol. Essential for product safety, especially in water-containing formulas.

Retinoid: A derivative of vitamin A. Includes prescription tretinoin and over-the-counter retinol, retinaldehyde, and retinyl esters. Increases cell turnover, stimulates collagen production, and addresses multiple signs of aging.

Rinse-Off Product: A formulation designed to be washed off after application, such as cleansers or masks.

Sebum: An oily substance secreted by the sebaceous glands in the skin. It is a natural moisturizer, but overproduction can contribute to acne.

Sensitization: An allergic response that develops after exposure. Re-exposure can trigger reactions at low levels.

Sensitizer: An ingredient that can cause allergic reactions or irritation, particularly with repeated exposure. Common sensitizers include fragrance, essential oils, and certain preservatives.

Silicone: Synthetic polymers used for smooth texture and occlusive properties. Prized for slip and barrier function. Examples include dimethicone and cyclopentasiloxane.

Solubilizer: A surfactant system used to disperse small amounts of oil or fragrance in water-based products.

Solvent: A liquid that dissolves other ingredients. Water is the most common solvent in skincare. Others include glycerin, propanediol, and various alcohols.

SPF (Sun Protection Factor): A UVB protection rating indicating how much longer skin can be exposed before sunburn versus unprotected skin under test conditions.

Stabilizer: An ingredient that keeps a formula consistent and prevents separation, oxidation, or degradation. Examples include tocopherol (vitamin E), ferulic acid, and various polymers.

Stratum Corneum: The outermost layer of the epidermis, consisting of dead skin cells held together by lipids. This "brick and mortar" structure is vital for barrier function.

Surfactant (Surface Active Agent): An ingredient that reduces surface tension between substances, allowing them to mix. In cleansers, surfactants lift oil and dirt from skin. Classes include anionic, cationic, amphoteric, and nonionic and range from gentle to harsh on skin.

Synergy: When two or more ingredients work together to produce an effect greater than the sum of their individual effects.

Synthetic Ingredient: Lab-created compounds that may mimic natural substances or provide enhanced stability.

TEWL (Transepidermal Water Loss): The amount of water that evaporates through the skin. High TEWL indicates barrier damage. Occlusives and barrier-repair ingredients reduce TEWL.

Toner: A liquid product used post-cleansing to balance pH, prep skin for treatment, remove residual impurities, and often deliver light hydration or mild exfoliation.

Tyrosinase: The key enzyme in melanogenesis (melanin production). Targeted by pigment-modulating ingredients.

UVA/UVB (Ultraviolet Radiation): Electromagnetic energy from the sun. UVA rays (longer wavelength) cause aging and long-term damage; UVB rays (shorter wavelength) cause sunburn. Both require sunscreen protection.

UV Filter (Organic/Chemical): Carbon-based UV absorbers that convert UV radiation to heat. Photostability depends on the system.

Visible Light: Light wavelengths visible to the human eye (approximately 400-700 nm). Can trigger hyperpigmentation in melanin-rich or melasma-prone skin. Iron oxide pigments help protect against visible light exposure.

Viscosity: The thickness or flow of a product, influenced by formulation and emulsifiers.

Volatile Silicone: A fast-evaporating silicone that provides slip and a dry finish.
W/O (Water-in-Oil) Emulsion: Oil-continuous emulsion with occlusive feel and higher water resistance. Common in rich creams and some sunscreens.

White Cast: A pale film left by some mineral sunscreens due to pigment scattering. Reduced by tinting with iron oxides and good dispersion

INDEX

ABOUT THE AUTHOR

Barbara Atkinson is a regenerative aesthetics professional dedicated to advancing longevity, skin health, and science literacy. Through her work, she helps people navigate the fast-changing landscape of regenerative treatments and evidence-based skincare. SKIN WISE extends this mission, providing readers with the same practical, science-grounded education she provides in her practice.

Barbara has witnessed firsthand how rapidly the world of skincare is evolving, from peptides and growth factors to exosomes, plant-derived nanovesicles, and polynucleotides, and she is committed to helping people separate genuine breakthroughs from marketing spin. She is passionate about empowering people to make smarter, more confident skincare choices so they can enjoy their best skin ever.

www.ingramcontent.com/pod-product-compliance
Lightning Source LLC
LaVergne TN
LVHW010947110826
845149LV00015B/3240

* 9 7 9 8 9 9 3 0 6 8 0 1 5 *